Chem 212
M. Sander
Fall '81
#81908443

Heidelberg
Science
Library

M. F. O'Dwyer
J. E. Kent
R. D. Brown

Valency

Springer-Verlag
New York
Heidelberg
Berlin

M. F. O'Dwyer
J. E. Kent
R. D. Brown
Department of Chemistry
Monash University
Wellington Road
Clayton, Victoria
Australia 3168

Library of Congress Cataloging in Publication Data

O'Dwyer, Michael Francis, 1930–
Valency.

(Heidelberg science library)
Includes index.
1. Valence (Theoretical chemistry) I. Kent, Jay E., 1939– joint author. II. Brown, Ronald Drayton, 1927– joint author. III. Title. IV. Series
QD469.038 1977 541'.224 77-8366

Printed in the United States of America.

9 8 7 6 5 4 3 2 1

ISBN 0-387-90268-6 Springer-Verlag New York

ISBN 3-540-90268-6 Springer-Verlag Berlin Heidelberg

Preface

This book was written after one of us published two editions of a smaller book (*Atomic Structure and Valency* by R. D. Brown, 1961 and 1966) and from the experiences of all three of us teaching a first-year university course in valency at Monash University. Our object has been to give enough of an historical introduction to quantum mechanics to enable a student to grasp the fundamental ideas without being subjected to much mathematical formalism. We have also tried to avoid making erroneous statements in the interest of simplicity (e.g., the widespread tendency to ignore the difference between $2p_1$, $2p_0$, $2p_{-1}$ and $2p_x$, $2p_y$, $2p_z$) because these lead to irritation and confusion of the better students, when they proceed to further studies of chemical quantum mechanics. The topics we have chosen to expand upon—i.e., energy levels of electrons in atoms, energetic considerations of bonding in diatomic molecules, and packing of ions in the simplest solid state structures—we believe form a good basis for students to progress to more complicated systems in a qualitative way. Both space and the intended level of the book have necessitated that the experimental section on spectroscopic and diffraction methods be very introductory. Again we believe that it is essential for a student to have some acquaintance with this, if the whole subject is to have a firm basis.

Acknowledgments

Criticism from our students, friends and associates, who have studied or read the manuscript is very much appreciated. We would also like to thank and acknowledge the untiring efforts and patience of those who helped us prepare the manuscript and artwork, Mrs. Edna Peebles, Mrs. Sharon Lampkin, Mr. Ben Baxter and his associates, and Mr. Don Ling.

Contents

1 Introduction

Before we embark on this introductory study of various theories and concepts about atoms and molecules, it is necessary to lay some groundwork concerning units of measurement of physical quantities. In particular, we are very interested here in the units of mass and energy. The units of mass are simple; chemists and, in fact, most scientists use the metric *gram*—or more correctly, as we see in a moment—the *kilogram*. With energy the common units are more varied among scientists and thus the student must take special care to avoid confusion. One of the best ways to avoid confusion in any unit is to define a consistent and standardized set of units. This was devised and internationally accepted in 1960; the set is known as *Système International d'Unités,* or simply SI units.

SI Units

These are *rational* or *coherent* units in that there are no constants built into the units. Known also as MKSA units, there are seven basic quantities from which all other units are derived. These seven units are given in Table 1.1.

Derived SI Units

All the other units of physical quantities derive directly from these seven basic SI units. Some of these have familiar names and symbols often named after famous scientists; those most frequently encountered are shown in Table 1.2.

Others do not have special names. Table 1.3 lists some examples of SI units in this category.

Table 1.1 Basic SI units

Physical quantity	Name of unit	Symbol for unit
Length	meter	m
Mass	kilogram	kg
Time	second	s
Electric current	ampere	A
Thermodynamic temperature	kelvin	K
Luminous intensity	candela	cd
Amount of substance	mole	mol

Table 1.2 Derived SI units with special names

Physical quantity	Name of unit	Symbol for unit	Definition of unit
Energy	joule	J	$kg\ m^2s^{-2}$
Force	newton	N	$kg\ m\ s^{-2} = J\ m^{-1}$
Power	watt	W	$kg\ m^2s^{-3} = J\ s^{-1}$
Pressure	pascal	Pa	$kg\ m^{-1}s^{-2} = N\ m^{-2}$
Electric charge	coulomb	C	A s
Electric potential difference	volt	V	$kg\ m^2s^{-3}A^{-1} = J\ A^{-1}s^{-1}$
Electric resistance	ohm	Ω	$kg\ m^2s^{-3}A^{-2} = V\ A^{-1}$
Electric capacitance	farad	F	$A^2s^4kg^{-1}m^{-2} = A\ s\ V^{-1}$
Magnetic flux	weber	Wb	$kg\ m^2s^{-2}A^{-1} = V\ s$
Inductance	henry	H	$kg\ m^2s^{-2}A^{-2} = V\ s\ A^{-1}$
Magnetic flux density	tesla	T	$kg\ s^{-2}\ A^{-1} = V\ s\ m^{-2}$
Luminous flux	lumen	lm	cd sr
Illumination	lux	lx	cd sr m^{-2}
Frequency	hertz	Hz	cycle per second : s^{-1}
Customary temperature, *t*	degree Celsius	°C	t [°C] = T [K] − 273.15
Plane angle	radian	rad	
Solid angle	steradian	sr	

Table 1.3 Derived SI units without special names

Physical quantity	SI Unit	Symbol for unit
Area	square meter	m^2
Volume	cubic meter	m^3
Density	kilogram per cubic meter	$kg\ m^{-3}$
Velocity	meter per second	$m\ s^{-1}$
Molar mass	kilogram per mole	$kg\ mol^{-1}$

Table 1.4 Fractions and multiples of SI units

Fraction	Prefix	Symbol	Multiple	Prefix	Symbol
10^{-1}	deci	d	10	deka	da
10^{-2}	centi	c	10^2	hecto	h
10^{-3}	milli	m	10^3	kilo	k
10^{-6}	micro	μ	10^6	mega	M
10^{-9}	nano	n	10^9	giga	G
10^{-12}	pico	p	10^{12}	tera	T
10^{-15}	femto	f	10^{15}	peta	P
10^{-18}	atto	a	10^{18}	exa	E

Prefixes for Fractions and Multiples of SI Units

Table 1.4 lists the internationally accepted prefixes for fractions and multiple of SI units.

These conform with the usual prefixes commonly used by scientists so that, for example, a frequency of 3.31×10^{10} Hz could equally well be expressed as 33.1 GHz (gigahertz) and 10^{-2} meter could be expressed as 1 cm (centimeter).

One awkward and as yet unresolved problem is that the basic SI unit for mass, the kilogram, already possesses a prefix. It is probably best to express 10^3 kg simply as one thousand kilograms or a megagram rather than a kilokilogram. Similarly, 10^{-3} kg is conveniently expressed as a gram rather than the absurd-sounding millikilogram!

Compound prefixes are taboo; for example, 10^{-9} meter is represented by

1 nm, *not* 1 mμm.

The attaching of a prefix to a unit in effect constitutes a new unit:

$$1 \text{ km}^2 = 1 \text{ (km)}^2 = 10^6 \text{ m}^2$$

and not

$$1 \text{ k(m)}^2 = 10^3 \text{ m}^2$$

Units Contrary to SI

You will encounter many units in various textbooks that are contrary to the SI rules. A few of these and their SI equivalents are given in Table 1.5.

In this book we endeavor to avoid use of units that are contrary to the SI codification such as those in Table 1.5. However, there are a few units very closely related to SI for which we make exceptions. Specifically, these few are listed in Table 1.6.

A table of physical constants is included for convenience inside the front cover.

Energy Units in Valency

Notice that there are two additional units of energy[1] in Table 1.6 aside from the SI joule. This is perhaps unfortunate, but nonetheless a fact. Chemists who classify themselves as having some degree of expertise in the field of valency are as much to blame for this state of affairs as anyone. The electron volt is the kinetic energy gained by an electron that has been accelerated through a potential difference of one volt. The electronvolt (eV) is simply the electronic charge e (see Chapter 2) times this potential difference V.

We have relaxed the SI principle in this book for the energy units of eV and cm^{-1} because not only are they all-pervading in other books and scientific articles, but also the magnitudes of the energy quantities are so minute when expressed in joules (we normally encounter molecular energies less than 1 aJ!). To assist in converting from one energy unit to another, Figure 1.1 gives a handy reference guide.

[1]Frequency is proportional to energy via Planck's constant, $E = h\nu$, where E is the energy in joules, h is Planck's constant ($h = 6.626 \times 10^{-34}$ J s), and ν (Greek nu) is the frequency in Hertz (s^{-1}). The development of this important relationship between energy and frequency is further depicted in Chapter 3.

Table 1.5 Units contrary to SI

Physical quantity	Contrary unit	SI Equivalent
Length	Å	10^{-10} m = 100 pm
Energy	erg	10^{-7} J
	thermochemical calorie	4.184 J
Force	dyne	10^{-5} N
	kilogram-force	9.80665 N
Pressure	atmosphere	101.325 kPa
	torr—mm Hg	133.322 Pa
Magnetic flux density	gauss	10^{-4} T

Table 1.6 Units to be allowed in conjunction with SI in this book

Physical quantity	Name of unit	Symbol for unit	Definition of unit
Energy	Electronvolt	eV	1.602×10^{-19} J
Wavenumber	Centimeter to the minus one or loosely referred to as wavenumbers	cm^{-1}	100 m^{-1}

Multiply by	from →	to	Multiply by
1.240×10^{-4}	eV molecule^{-1}	cm^{-1} molecule^{-1}	8065.5
5.034×10^{22}	cm^{-1} molecule^{-1}	joule molecule^{-1}	1.986×10^{-23}
6.242×10^{18}	eV molecule^{-1}	joule molecule^{-1}	1.602×10^{-19}
2.390×10^{-4}	kcal mol^{-1}	joule mol^{-1}	4.184×10^{3}
Multiply by	to ←	from	

FIGURE 1.1. Energy conversion.

Classical Forms of Energy

Before leaving the topic of units and energy, it may be useful to review briefly a general classical classification of energy since much of our subsequent discussions are directly concerned with the energy of atoms and molecules. The *total mechanical energy* of a body is defined as the sum of its potential and kinetic energies.

$$E_{\text{tot}} = E_{\text{pot}} + E_{\text{kin}} = mgh + \tfrac{1}{2}mv^2$$

m being the mass of the object, g the Earth's gravitational constant = 9.80 m s^{-1}, h the height above a reference plane, and v the body's velocity. In addition to potential energy arising from gravity ($E_{\text{pot}} = mgh$), other forms of potential energy can arise. The one that we encounter most frequently in valency is electrostatic potential energy:

$$E_{\text{pot}} = \varphi Q$$

where Q is the electric charge on the object and φ is the electrostatic potential experienced by the object (e.g., when the object is a distance r meter from a charge q coulomb, it experiences a potential of $\varphi = (q/4\pi\epsilon_0 r)$ V).[2]

In addition, energy manifests itself thermally in terms of heat energy. The general study of the relations between other forms of energy and heat is known as thermodynamics. At the level that we discuss atomic structure and valency here we do not need to delve deeply into this subject but we use an energy principle:

> The most stable state of an individual atom or molecule is that of lowest total potential energy. When there are external influences, such as collisions with other molecules or radiation, the atom or molecule may be found in some

[2] $1/4\pi\epsilon_0$ is a proportionality constant, ϵ_0 referred to as the *permittivity of vacuum* and having the value $\epsilon_0 = 8.855 \times 10^{-12}$ s^{2}C^2 kg^{-1} m^{-3}.

> other energy state. Under normal laboratory conditions most atoms and molecules are encountered in their most stable state.

This is a principle where observation constitutes the proof, and for our purposes we accept it as being true.

Finally we also should realize that classically the transmission of energy can occur through the use of: (a) *projectiles* where mass is transferred and carries energy obtained at point a to point b and (b) *waves* where mass is not transferred.

An example of the former is a bowler who exerts himself to give kinetic energy to a cricket ball that then travels down the pitch carrying its energy (primarily kinetic energy) with it. The energy is expended on impact of the ball with the wicket, the bat, the stumps, or the gloves of the wicket keeper. Thus the ball acts as the projectile that transfers the energy. Transfer of energy via wave motion is more subtle in that the transportation of matter does not take place. For example, an earthquake in New Zealand causing a tidal wave to hit the eastern coast of Australia does not mean that water from New Zealand has been hurled against the shores of Australia, but simply that the ocean has been disturbed in one place, the disturbance has traveled across the surface of the water, and finally caused damage on the beaches of another shore. Other examples are seismic waves (earthquakes) and the cracking of a stock whip.

The two ways of classically transmitting energy historically were understood with the exception of one phenomenon known as light. Light is an energy form, but its propagation could not be explained using *either* a projectile model or wave model, and classical arguments appear to break down. As we see shortly, the significance of this is quite profound and has far-reaching implications in the currently accepted theories of atomic structure and valency.

Problems

1.1 Find the dimensions of action (energy $\times$ time).

1.2 For the ideal gas equation, $PV = nRT$, find the SI units of R.

1.3 Express the weber units (see Table 1.2) in terms of J (joules) and A (amperes).

1.4 Complete the following equations:
a. $J \div C =$
b. $C \times \Omega = \mathrm{V} \times$
c. $J \div N =$

1.5 Express the following:
a. 2.562×10^{-7} m in nm
b. 5685 g as kg
c. 1.09×10^{-2} kV as mV
d. 2.07×10^{-21} J as aJ

1.6 The electric permittivity of vacuum is defined by:

$$\epsilon_0 = \frac{1}{\mu_0 c^2}$$

where μ_0 is the magnetic permeability of vacuum and is by definition equal to $4\pi \times 10^{-7}$ H m^{-1}. Show that the units of ϵ_0 are s^{2}C^2 kg^{-1} m^{-3}.

1.7 What is the conversion factor for eV molecule^{-1} to J mol^{-1}?

2 Gross Atomic Structure

To begin our study of atomic structure and valency, we must ask a question that occurred to the earliest philosophers. If we examine a material body that appears homogeneous and then divide it into successively smaller pieces, can the division process go on indefinitely or will we eventually find a building block of matter that cannot be further subdivided without losing the characteristics of the material body? The answer, as postulated by Democritus (ca. 460 B.C.) and John Dalton (1766–1844), for example, is an emphatic "no." We eventually reach an *atomic* or *molecular* state in which further degradation does alter the basic properties of our material.

We do not spend time here recounting the evidence (such as the proportions in which atoms combine to form compounds) that led to the atomic hypothesis. Rather, we trace the early development of the understanding of the gross structure of the atom from an experimental viewpoint. In Chapter 3 the explanations and theories of atomic structure are developed.

The Early Clues

It is difficult to impart to students in the 1970s the excitement and challenges that the early workers must have felt as they pieced together the structure of the atom. Facts that are taught nonchalantly today and accepted in the same manner defeated the world's greatest intellects for centuries.

The first really significant insight was gained through Michael Faraday's experiments on the electrolysis of substances in solution in the late nineteenth century. These experiments indicated that electrical phenomena and matter were in some way related. From this, the concept of an indivisible *unit of electricity* was proposed and named the *electron* in 1874 by G. Johnstone Stoney.

These electrochemical experiments, showing that matter possesses charge, were supplemented by the study of electrical discharges in gases carried out primarily in the English laboratories of William Crookes and J. J. Thomson. A discharge tube (Figure 2.1) filled with 100 kPa pressure ($\approx$1 atmosphere) of gas requires an electric field of some 4 MV m^{-1} before any measurable current will pass through the gas. At this high field a spark is produced.

If we now pump on the discharge tube and remove most of the gas, the spark appears to broaden out as the pressure is lowered and becomes a uniform luminous glow within the tube at a pressure of about 0.5 Pa. The glow depends not only on the pressure, but also on the type of gas present and the dimensions of the tube. Neon signs used in advertising are familiar examples of glow-discharge tubes of this type filled with neon gas. With the meter M we can easily detect a current of electricity passing through the gas.

It was postulated that the current is carried between the two electrodes of the discharge tube by electrons and gaseous ions. Positive ions are produced by removal of one or more electrons from the gas atom or molecule, and negative ions are produced should an electron become attached to the atom or molecule. The electrons are present to some extent before any current begins to flow and upon acquiring sufficient velocity in the electric field can collide with the gaseous atoms, causing ionization by knocking off further electrons, which in turn causes still

FIGURE 2.1. A gas-discharge tube.

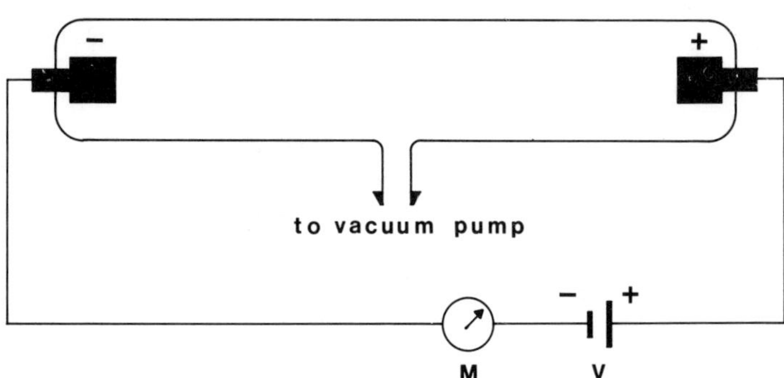

further ionization as the ions and electrons migrate toward the charged electrodes. (The glow of the tube is incidental to the above argument and is caused by excessive excitation of the ions produced, which then relax, emitting visible light in the process.) The main properties of the electron were determined by two ingenious experiments, one by Thomson measuring the ratio of mass to charge of the electron and another by R. A. Millikan, measuring the charge of the electron (known as the classical "oil-drop experiment"). The modern value of the electronic charge is expressed in coulombs as:

$$-e = -1.602 \times 10^{-19} \text{ C}.$$

From this value the mass from the m/e value of Thomson comes out to be:

$$m_e = 9.109 \times 10^{-31} \text{ kg}$$

without considering relativity effects. That is, this value represents the so-called rest mass of the electron. So it would appear from this that a fair bit was known about the electron, at least insofar as it was part of the atom. What about the rest?

Here the clue to the gross structure of the atom came from the laboratory of Sir Ernest Rutherford, an expatriate New Zealander. His work, which we describe shortly, was based on the phenomenon of radioactivity discovered by Henri Becquerel in 1896 in minerals that we now know to contain the heavy element uranium. From an experiment whereby a lump of uranium brought near an electroscope caused the electroscope to discharge, Becquerel correctly deduced that the uranium mineral was emitting some sort of radiation that ionized the air. Subsequently Pierre and Marie Curie found other elements that also possessed this property, namely radium and polonium.

The ionizing property of these radioactive substances was demonstrated to be due to three types of radiation: (a) high-speed electrons known as β-rays or β-particles, (b) particles of a mass of approximately 7300 times the rest mass of the electron and carrying a positive charge twice that of an electron (termed α-particles by Rutherford), and (c) a type of radiation not deflected by magnetic or electric fields and similar in property to light but of much higher degrees of energy and penetration (known as γ-rays).

Knowledge that atoms contain electrons and are capable of emitting radiation paved the way for Rutherford's experiments and model for the atom.

The Rutherford Experiment

Rutherford used the α-particles emitted by radioactive substances as probes to investigate the internal structure of atoms. He reasoned that since atoms were known to consist partially of negatively charged electrons, they must also carry some positively charged constituent since the atoms in general were neutral.

α-Particles carry a positive charge; thus electrostatic forces should be present when the α-particles come near other charged particles such as those that make up the atom. Rutherford's experiment is shown schematically in Figure 2.2.

A small radioactive source such as a chunk of polonium emits α-particles, which are beamed through a slit onto a very thin (100 nm) piece of gold foil through which the α-particles can penetrate. The detector, such as a photographic plate or scintillating screen, tells us whether the α-particles come through the foil unhindered or are electrostatically deflected through some angle θ.

The results of the experiment can be summarized as follows: Most of the α-particles whizzed through the foil unhindered. A few α-particles were deflected by varying angles up to 180°. A simple calculation shows the gold foil to be about 400 gold atoms thick; hence Rutherford correctly deduced the atom was primarily empty space with a small massive central core or nucleus from which a few α-particles (1:10,000) were deflected by some force.

This strong force is simply a coulombic force of repulsion due to the positively charged α-particles interacting with the small massive positive nucleus of the gold atom. The expression for Coulomb's law is

$$F = k\left(\frac{q_1 q_2}{r^2}\right) \tag{2.1}$$

FIGURE 2.2. The Rutherford experiment.

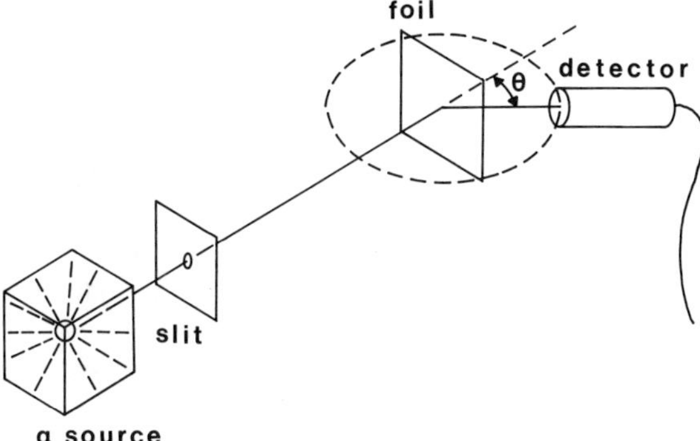

where k is a proportionality constant and where the force is proportional to the magnitude of the two charges q_1 and q_2 and inversely proportional to the square of the distance between the two charges. From this we can find the dimensions of the atom in much the same way Rutherford and his students did. The potential energy in turn is expressed as (see Chapter 1):

$$E_{\text{pot}} = \frac{q_1 q_2}{4\pi\epsilon_0 r} \tag{2.2}$$

where here we have put in the constant of proportionality $1/4\pi\epsilon_0$.[1] Now the α-particle has a charge of $+2e$ and the gold atom nucleus of $+Ze$, Z being the atomic number. Thus the potential energy at some distance a is:

$$E_{\text{pot}} = \frac{2Ze^2}{4\pi\epsilon_0 a} \tag{2.3}$$

If we imagine the α-particle to be aimed directly at the gold nucleus, it will ricochet back at an angle of 180° and at some instant, when the α-particle is at rest, all of its kinetic energy will be converted to potential energy given by Equation 2.3. The α-particle emanating from a radioactive source has kinetic energy corresponding to about 4×10^6 eV (6.4×10^{-13} J). Equating this to the potential energy at distance a we have:

$$\begin{aligned} a &= \frac{Ze^2}{2\pi\epsilon_0}(E_{\text{pot}}) \\ &= Z \times \frac{(1.602 \times 10^{-19})^2}{6.283 \times 8.855 \times 10^{-12} \times 6.4 \times 10^{-13}} \end{aligned}$$

If the foil is gold, $Z = 79$, and we find that a is about 6×10^{-14} m. Of course in reality, near misses by the α-particle are much more probable than direct hits. Furthermore, as previously mentioned, complete misses occur most frequently. Nonetheless, by extensions of the above arguments we can calculate the probability of any ricochet angle of the α-particles from the experimental evidence of Rutherford if we assume: (a) the nuclei to be very small with respect to the atoms themselves and (b) an inverse square law of repulsive force (Coulomb's law) to exist between the positively charged nuclei and the positively charged α-particles.

The agreement between such calculations and the experimental observations is very good and therefore we

[1] $\epsilon_0 = 8.855 \times 10^{-12}$ s^{2}C^2 kg^{-1} m^{-3}.

can set an upper limit to the size of the nuclei of about 10^{-14}m.

To summarize, the Rutherford atom model consists of a nucleus of about 10^{-14} to 10^{-15} m radius carrying a charge of $+Ze$ surrounded by Z moving electrons each with a charge of $-e$. We do not know as yet from this model what gives the atom stability, but the stage is set with the Rutherford atom model for the development of atomic theory that took place early in the present century.

Discovery of the Neutron and Periodic Classification of the Elements

The elementary nuclear particle carrying a positive charge discovered by Rutherford is called the proton. We have already used the relationship that the number of protons, p, is related to the atomic property Z:

$$p = Z.$$

In 1932 Sir James Chadwick showed that, in addition to protons, atomic nuclei also contained particles of similar mass to the proton but uncharged. He called them *neutrons*. The masses of atoms are easily obtained using an instrument known as a mass spectrometer, and then by knowing the electron mass presumably we can deduce the mass of the proton and neutron. However, the picture is not quite this simple, since various isotopes (elements with the same Z but differing number of neutrons) are present in natural abundance for most elements.[2]

The accurate rest masses for the proton and neutron are

$$m_{\mathrm{p}} = 1.6726 \times 10^{-27}\ \mathrm{kg}$$
$$m_{\mathrm{n}} = 1.6749 \times 10^{-27}\ \mathrm{kg}.$$

Both nuclear particles are very similar in mass and about 1840 times more massive than the electron.

It was recognized early in the game that a relative scale of these small masses would be more useful. Since 1961 this relative scale has been based on an isotope of carbon containing six protons and six neutrons and given the relative mass

$$^{12}\mathrm{C} = 12.000000\ u.$$

We define a *unified atomic mass constant, u,* which is one-twelfth the mass of a single atom of carbon-12; $u = 1.66043 \pm 0.00008 \times 10^{-27}$ kg. On this scale the proton mass and neutron mass become very nearly equal to unity:

[2]In addition, mass manifests itself as nuclear binding energy via the Einstein relation $E = mc^2$. Thus adding the masses of the protons, neutrons, and electrons does not give the exact atomic mass.

Table 2.1 Properties of some fundamental atomic particles

Particle	Symbol	Mass [u]	Mass [kg]	Charge [C]
Proton	p	1.00727	1.6726×10^{-27}	$+1.602 \times 10^{-19}$
Neutron	n	1.00867	1.6749×10^{-27}	0
Electron	e	0.000549	9.1096×10^{-31}	-1.602×10^{-19}

$$m_{\mathrm{p}} = 1.00728\ u$$
$$m_{\mathrm{n}} = 1.00867\ u.$$

Table 2.1 summarizes the properties thus far of our atomic particles.

Hence we are able to define a mass number M of a nucleus as the mass to the nearest integer on our atomic mass scale

$$p + n = M$$

where p is the number of protons and n represents the number of neutrons. With this information the composition of any nucleus is readily determined. For example, $^{19}_{9}\mathrm{F}$, the common isotope of fluorine, has

$$p = 9, \qquad p + n = 19 \quad \text{hence } n = 10$$

that is, there are nine protons and 10 neutrons in the nucleus. The actual mass of the $^{19}_{9}\mathrm{F}$ atom is 18.9984 u.

Atomic weights (more correctly, masses) used by chemists are weighted averages of the individual masses of the atoms present in a particular element in natural abundance. They are expressed in unified atomic mass units and thus reflect the proportionality constant between the real atomic mass scale (in kg) and our relative atomic mass scale (in u) based on the carbon-12 isotope:

$$1\ \mathrm{kg} = 6.0222 \times 10^{26}\ u.$$

We say "reflect" because we further define for calculational convenience a *mole* (mol), which contains the same number of atoms (or ions, electrons, molecules, as the case may be) as there are atoms in exactly 0.012 kg of carbon-12. This number of atoms is, of course, Avogadro's number and has the value:

$$N_A = 6.02222 \times 10^{23}\ \mathrm{mol}^{-1}.$$

Atomic or molecular weights are often seen expressed as the *molar mass,* which has the units of kg mol^{-1}. For example, the molar mass of carbon taking into account the 1.1% natural abundance of carbon-13 is 0.01201 kg mol^{-1}. Notwithstanding our SI units, it is more frequently given as 12.01 g mol^{-1}, so that the numerical value of the atomic weight in u and the molar mass are the same. Table 2.2 lists the masses of a few of the lighter elements.

Before closing this chapter there are two very important points we wish to make; these are the principal topics treated in the ensuing chapters. Firstly, when an atom is united with others to form a molecule, it is found that the properties of the nucleus are unaltered from those of the free atom, but the same is *not true* of the electrons moving around the nucleus. Many of the properties of the electrons in an atom, such as their energies, modes of motion, and magnetic properties are profoundly altered when an atom unites chemically with others. From this it is evident that our development of theories of valency will be primarily based on a knowledge of the electronic structure of atoms rather than the nuclear structures. Secondly, although Rutherford's scattering experiments provided sound evidence for the validity of the nuclear model of the atom, it was most strongly in conflict with the accepted classical theories of electromagnetism. The basis of this conflict was the important question of the *stability* of the nuclear atom itself. On the one hand, if the electrons were *statically* placed around the nucleus, the atom could not be in an equilibrium state because of the known electric forces between point charges[3] and on the other hand, if the electrons were orbiting the nucleus in a *dynamic* fashion the classical laws of electromagnetism predicted the atom would necessarily radiate energy and rapidly collapse.

Clearly a radically new approach was needed to explain atomic stability. This was forthcoming with the development of the *quantum-mechanical* model of the atom, which is described together with its consequences in Chapters 3 and 4. In Chapter 5 we introduce the subject of valency to show how our stable quantum-mechanical atoms can become "social" and combine to form molecules.

[3]A theorem in electrostatics states that no stationary arrangement of point charges is stable.

Table 2.2 Masses of some of the lighter elements

Z value	Element	M	Atomic mass [u]	Natural abundance [percent]	Atomic weight [u]
1	Hydrogen (H)	1	1.007825	99.985	1.00797
		2	2.01410	0.015	
		3	ND[a]		
2	Helium (He)	3	3.01603	0.00013	4.0026
		4	4.00260	99.99987	
3	Lithium (Li)	6	6.01513	7.42	6.939
		7	7.01601	92.58	
4	Beryllium (Be)	7	ND[a]		9.0122
		9	9.01219	100	
		10	ND[a]		
5	Boron (B)	10	10.01294	19.6	10.811
		11	11.00931	80.4	
6	Carbon (C)	12	12.00000	98.89	12.01115
		13	13.00335	1.11	
		14	ND[a]		
7	Nitrogen (N)	14	14.00307	99.63	14.0067
		15	15.00011	0.37	
8	Oxygen (O)	16	15.99491	99.759	15.9994
		17	16.99914	0.037	
		18	17.99916	0.204	
9	Fluorine (F)	19	18.99840	100	18.9984
10	Neon (Ne)	20	19.99244	90.92	20.183
		21	20.99395	0.257	
		22	21.99138	8.82	
11	Sodium (Na)	23	22.98977	100	22.9898
12	Magnesium (Mg)	24	23.98504	78.70	24.312
		25	24.98584	10.13	
		26	25.98259	11.17	
13	Aluminium (Al)	27	26.98153	100	26.9815
14	Silicon (Si)	28	27.97693	92.21	28.086
		29	28.97649	4.70	
		30	29.97376	3.09	
15	Phosphorus (P)	31	30.97376	100	30.9738
16	Sulfur (S)	32	31.97207	95.0	32.064
		33	32.97146	0.76	
		34	33.96786	4.22	
		36	35.96709	0.014	
17	Chlorine (Cl)	35	34.96885	75.53	35.453
		37	36.96590	24.47	
18	Argon (Ar)	36	35.96755	0.337	39.948
		38	37.96272	0.063	
		40	39.96238	99.60	

[a] ND—radioactive, mass not determinable.

Problems

2.1 What are the relative masses of the isotopes $^{35}_{17}Cl$, $^{12}_{6}C$, and $^{15}_{7}N$?

2.2 What are the molar masses of NaCl, CCl_4, benzene, and SO_2?

2.3 What is the charge, in C, on Avogadro's number of electrons?

3 Atomic Theory

We mentioned in Chapter 1 that the phenomenon of light was to present difficulties to scientists of the late nineteenth and early twentieth centuries. There appeared to be two ways of describing light, either in terms of a projectile or *particle model* (the particles of light are known as *photons*) as championed by Newton using Newtonian mechanics, or in terms of *classical wave motion,* which gained very solid support from diffraction experiments. The fact that light could be diffracted seemed clear evidence that it could be ascribed wavelike characteristics. The electromagnetic wave theory due to Maxwell was applied to the behavior of light very successfully. By the turn of the century the wave theory of light had gained almost universal acceptance.

In classical wave theory, light energy can be described in terms of wavelength, λ, in amplitude–distance coordinates such as in Figure 3.1(a), or in terms of frequency, ν, in amplitude–time coordinates as in Figure 3.1(b). The frequency of the light is related to the wavelength (λ) by the relationship

$$\nu = \frac{c}{\lambda} \tag{3.1}$$

where c is a universal constant equal to the velocity of light *in vacuo* and has the value

$$c = 2.998 \times 10^8 \text{ m s}^{-1}. \tag{3.2}$$

We will not worry that the velocity of light varies very slightly if the medium through which it travels is not a vacuum. A value of 3×10^8 m s^{-1} will usually be adequate.

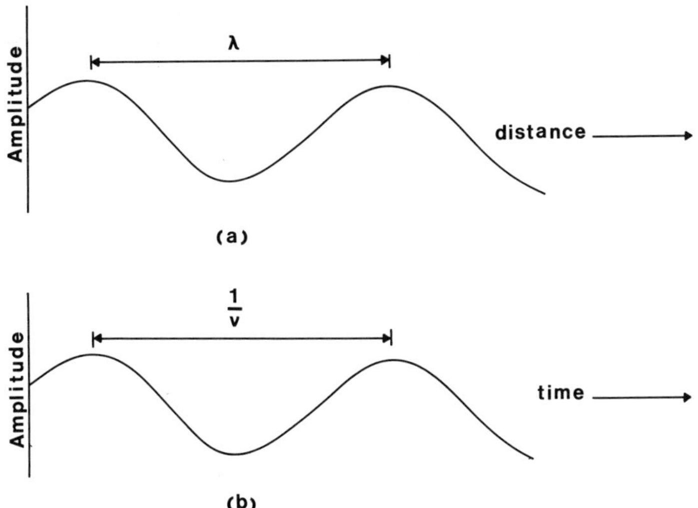

FIGURE 3.1. Representation of wave motion showing: (a) wavelength, λ; (b) frequency, ν.

As mentioned in Chapter 1, electromagnetic radiation is often specified by quoting its reciprocal wavelength (wavenumbers; cm^{-1}), which is proportional to frequency, as can be seen from Equation 3.1.

$$\tilde{\nu}(\mathrm{cm}^{-1}) = \frac{1}{\lambda(\mathrm{cm})}. \tag{3.3}$$

When speaking of light we in general conjure up visible light, at least that visible to the naked eye. Visible light, however, has a very narrow frequency range and is but one instance of a more general variety of radiation. The *electromagnetic spectrum* embraces all types of radiation from very short wavelength γ-rays to very long wavelength radiowaves. The electromagnetic spectrum is diagramatically represented in Figure 3.2. The spectrum is broken up into regions, each region having a name, such as infrared, visible, ultraviolet, and so on. All electromagnetic radiation travels at the same speed and obeys the wavelength:frequency relationship given by Equation 3.1.

Two Spanners in the Works

Even though we commonly use the wave-like properties just mentioned to describe light and other forms of electromagnetic radiation, two critical experiments have cast some doubt on this classical description.

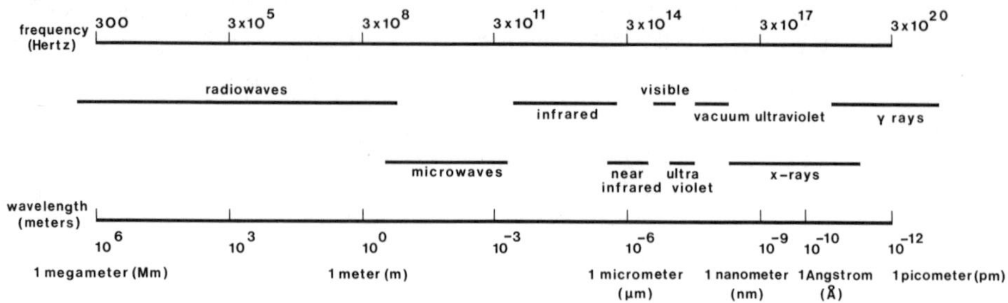

FIGURE 3.2. The electromagnetic spectrum.

The Michelson–Morley Experiment

All wave motion, whether from sound waves or waves created by pebbles tossed in a pond, depends on some medium—air or water in these cases—for propagation. The fact that electromagnetic radiation could be described with wavelike characteristics naturally led early investigators to speculate as to the exact propagating medium for light waves. The nineteenth-century physicists coined the term *luminiferous ether* to describe this medium, which necessarily had to have the properties of vanishingly small density and viscosity since all bodies be they planets or feathers seemed to move through it without hindrance.

In 1881 A. A. Michelson devised a clever experiment sufficiently sensitive to detect the effect of motion of the equipment through the "ether" on the speed of light as measured by the equipment so that the existence of the "ether" could be detected. The experiment failed in the sense that the expected results did not occur; that is, there did not seem to be any "ether." Even with a more sensitive system, repetition of the experiment by A. A. Michelson and E. W. Morley in 1887 still could not reveal any effect due to the ether. The unavoidable conclusion that Michelson and Morley were forced to reach was that electromagnetic radiation, that is, light, was somehow different from other known types of wave phenomena. *Electromagnetic radiation does not need a medium for propagation.* So despite the fact that we seem to be able to describe electromagnetic radiation with wavelike characteristics, physicists at the beginning of the twentieth century had a nagging doubt as to whether they were really seeing the whole story. This doubt was increased by yet another inexplicable experimental result.

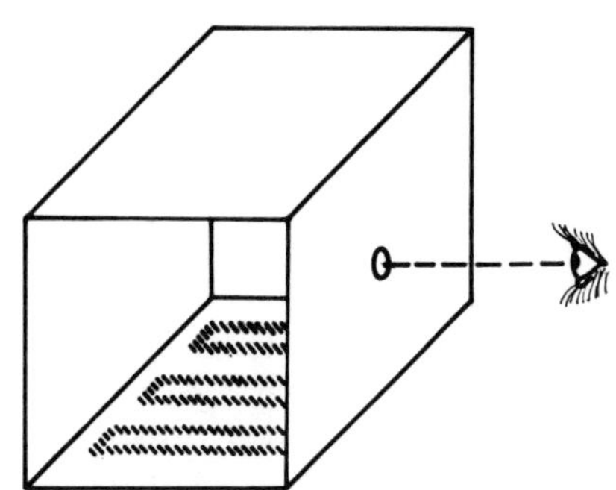

FIGURE 3.3.
A blackbody consisting of a hollow box heated electrically on the inside. Radiation inside box is viewed via a small hole in side of box.

Blackbody Radiation

When a material body is heated to a certain temperature we all know that it glows; hence the term "red-hot." Different materials but of similar size and shape will in general radiate visible light as they are heated, but it may be at different rates. Furthermore, the light may have a different spectral distribution for different materials.

If we now use a *hollow* body and view the radiation inside the body via a small pinhole in its wall, such as shown in Figure 3.3, it is found that for a given temperature the intensity and spectral distribution are the same for all materials regardless of the material's makeup or size. The phenomenon is known as blackbody radiation. Plots of intensity against wavelength curves at various temperatures are shown in Figure 3.4. It can be seen here that there are two features of interest.

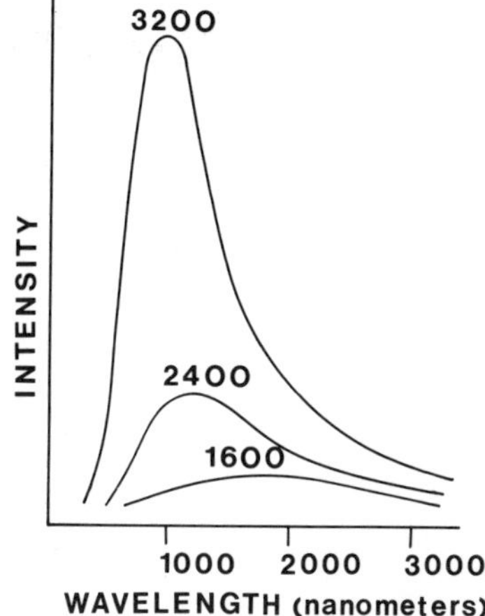

FIGURE 3.4.
Spectral distribution and intensity from a blackbody at different temperatures. (Temperatures in K shown on curves.)

1. The total amount of radiation (given by the area under the curves) increases rapidly with increasing temperatures.
2. The wavelength corresponding to the maximum in the spectral curve shifts to progressively shorter wavelength with increasing temperature.

Attempts to explain the temperature and intensity distribution of blackbody radiation using classical statistical mechanics and thermodynamics all failed, even though they went so far as to consider the nature of the atomic processes involved within the body. The reason that classical treatments failed is because they all assumed that the light emitted was continuous. It certainly *appeared* continuous from the spectrum because all wavelengths are represented even though they varied in intensity.

It was only when Max Planck in 1900 made the bold assumption that the radiation was not continuous, but consisted of tiny packets of energy that he referred to as *quanta,* that an explanation of blackbody radiation was possible. Specifically, he assumed that the cavity of the blackbody consisted of a large assembly of oscillating charges. These oscillators vibrated back and forth with characteristic frequencies and could absorb and emit energy, but only when the total energy absorbed or emitted was an exact integral multiple of the oscillators' vibrational frequency. In other words, the radiation emanating from the blackbody had an energy of:

$$E = nh\nu \tag{3.4}$$

where ν is the frequency of the oscillator (it could have any frequency; hence the continuous appearance of the spectra), n is an integer, and h is a characteristic constant known as Planck's constant:

$$h = 6.6262 \times 10^{-34} \text{ J s}. \tag{3.5}$$

Thus an oscillator of fundamental frequency ν, could radiate energy of $h\nu$, $2h\nu$, $3h\nu$, . . . , $nh\nu$, but could never radiate, say, 0.6 $h\nu$.

Planck's hypothesis of the *quantum theory of electromagnetic radiation* gives us the relationship between frequency and energy of light but more importantly it sets the stage for the *quantum theory of matter,* which is so crucial to our study of atomic theory and valency.

The student may well ask at this stage for a little more information about these quanta of light energy. For instance, how big are they? Here Planck's constant, h, tells us they are very small indeed, analogously to the manner in which Avogadro's number, N_A, tells us how small atoms are. In fact, if we combine the Einstein equation

$$E = mc^2 \tag{3.6}$$

with Planck's equation

$$E = h\nu \tag{3.7}$$

we can calculate the hypothetical[1] mass of any single quantum of light energy

$$h\nu = mc^2 \tag{3.8}$$

or

$$m = \frac{h\nu}{c^2} \tag{3.9}$$

since h (in J s) is a very small number and c^2 (in m^2s^{-2}) is very large, quite obviously a single quantum of light is indeed very light!

Similarly, the linear momentum p, which is just mass times velocity, can be calculated since light and other forms of electromagnetic radiation move at the velocity of light c.

$$p = mc \tag{3.10}$$

$$p = \frac{h\nu}{c^2}\,c \tag{3.11}$$

that is:

$$p = \frac{h\nu}{c} \tag{3.12}$$

Equation 3.12 can be expressed in terms of wavelength if we remember that $\lambda = c/\nu$:

$$p = \frac{h}{\lambda} \tag{3.13}$$

The linear momentum is also a small number (in kg m s^{-1}) because of the small value of h (in J s). Equation 3.13 is known as the *de Broglie relation.*

Dual Nature of Light and Matter

From Equation 3.13, which results from Planck's quantum hypothesis, it becomes apparent that description of light is dualistic—either particulate or wavelike descriptions can be given. That light has momentum, a particlelike prop-

[1]Actually, photons are massless; otherwise relativity theory predicts them to have infinite energy and momentum.

erty, was experimentally confirmed by A. H. Compton in 1922 using X-rays as his light. In the Compton effect the X-rays colliding with electrons act just like billiard balls.

Furthermore, the photoelectric effect whereby light impinged upon a metal surface caused, at certain wavelengths, electrons to be ejected from the metal was explained by Einstein by assuming the light was composed of quanta of light energy, namely, photons.

A further consequence of Equation 3.13 is that it applies not only to the phenomenon of light but also to matter proper. The equation says that all matter has wavelike properties—in other words, a characteristic wavelength given by the de Broglie relationship. The catch is that the value of the momentum of ordinary macroscopic particles is so large compared with Planck's constant h that the wavelength λ becomes negligibly small. For small submicroscopic particles such as our constituents of the atom, protons, neutrons, and particularly electrons, the wavelength may be discernible, thus nicely illustrating that the duality of nature applies not only to electromagnetic radiation but also to all matter. Let us perform just a couple of simple calculations to illustrate this duality concept.

The Wavelength of Sally Brown

Sally weighs 50.0 kg and moves nicely but slowly at a velocity of 0.250 m s^{-1}. Consequently, her momentum is

$$\begin{aligned} p = mv &= (50.0 \text{ kg})\,(0.250 \text{ m s}^{-1}) \\ &= 12.5 \text{ kg m s}^{-1} \end{aligned}$$

Her wavelength according to de Broglie's relationship is given by

$$\begin{aligned} \lambda = \frac{h}{p} &= \frac{6.63 \times 10^{-34} \text{ J s}}{12.5 \text{ kg m s}^{-1}} \\ &= 0.53 \times 10^{-34}\,(\text{kg m}^2\text{s}^{-1})\,(\text{kg}^{-1}\text{m}^{-1}\text{s}) \\ &= 5.3 \times 10^{-35} \text{ m.} \end{aligned}$$

So we see that although our gal Sal may be curvy, she has a negligible wavelength—beyond any possibility of detection.

The Wavelength of an Electron

Now let us consider an electron ($m = 9.1 \times 10^{-31}$ kg) moving with a velocity of 0.01 times c, the velocity of light. Its momentum is:

$$\begin{aligned} p = mv &= (9.1 \times 10^{-31} \text{ kg})(3 \times 10^{6} \text{ m s}^{-1}) \\ &= 27.3 \times 10^{-25} \text{ kg m s}^{-1} \end{aligned}$$

and its wavelength is:

$$\lambda = \frac{h}{p} = \frac{6.63 \times 10^{-34}\ \mathrm{J\ s}}{2.73 \times 10^{-24}\ \mathrm{kg\ m\ s^{-1}}}$$
$$= 2.4 \times 10^{-10}\ \mathrm{m} = 240\ \mathrm{pm}.$$

This may seem like a pretty small wavelength as well, but it is of the order of atomic dimensions and hence is detectable. How? Well, by diffraction techniques, for example. Diffraction of light is a wave characteristic so that demonstrating diffraction for electrons confirms the wavelike properties of this particle. The regular atomic spacings of a crystalline material should serve well as a diffraction grating (see Chapter 7 for more on diffraction), and in fact C. Davisson and L. H. Germer of the Bell Telephone Laboratories in the United States and also G. P. Thomson (son of J. J. Thomson) in England confirmed the diffraction of electrons by a nickel crystal in 1927, shortly after the de Broglie hypothesis appeared in 1925. The wave characteristics for particles of atomic dimensions have been frequently demonstrated since. Neutron diffraction experiments, for example, are carried out routinely in many laboratories as an excellent method of crystal analysis (again, see Chapter 7). Similarly, electron microscopes exploit the wave nature of electrons.

So duality is general and our classical theories of electromagnetism and mechanics simply did not hold for the submicroscopic world of atoms and molecules. New theories were needed. They came forth in the formalism of a new mechanics, specifically, quantum mechanics.

The Spectrum of Atomic Hydrogen

To develop the new mechanics of the quantum as a basis for atomic theory and ultimately, we hope, molecule formation, we again begin with experimental evidence known to the physicists of the late nineteenth and early twentieth centuries. This evidence had accumulated since Sir Isaac Newton in 1665 demonstrated that sunlight could produce a *spectrum* of different colors if first passed through a glass prism. A century and a half later Josef Fraunhofer, G. R. Kirchoff, Robert W. Bunsen (of burner fame), A. J. Angström, and others carried out extensive experimental work on the spectra of the sun and the elements. They were able to demonstrate that if an element absorbs light of a certain wavelength, then it will also emit the same wavelength when excited by a flame or discharge tube.

Because the sun, which contains a large percentage of hydrogen atoms, was such a convenient source of light emission, the spectrum of atomic hydrogen became very well known and received much attention. The spectrum in the visible region consists of a series of sharp lines called the *Balmer series,* (see Figure 3.5) that converge to a continuum at shorter wavelengths.

There are similar series of lines in the ultraviolet and infrared regions, and these are named after their discoverers—Lyman, Paschen, Brackett, and Pfund. The complete hydrogen-atom spectrum is shown in another representation in Figure 3.6 known as an *energy-level diagram* because of the relationship, clarified in the next few pages, between the spectral lines and the possible values of the energy of an electron in a hydrogen atom. The fact that we see discrete lines rather than a continuous range of wavelengths in the spectrum is directly related to the fact that only certain discrete values are possible for the energy of the atomic electron. The discreteness again is similar to the postulate of discreteness given by Planck for blackbody emission.

Notice in Figure 3.6 the integer n relating the various series of lines in the hydrogen-atom spectrum. This relationship was recognized early in the study by J. J. Balmer and J. R. Rydberg. Balmer first gave a formula for the Balmer lines. In wavenumbers this is:

$$\tilde{\nu} = R\left(\frac{1}{4} - \frac{1}{n^2}\right) \tag{3.14}$$

where n is an integer greater than 2 and R is a universal constant known as *Rydberg's constant.*

$$R = 109737.31\ \text{cm}^{-1} = 3.28985 \times 10^{15}\ \text{Hz}. \tag{3.15}$$

FIGURE 3.5. The emission spectrum of atomic hydrogen in the visible region of the electromagnetic spectrum. These sharp lines are known as the Balmer series.

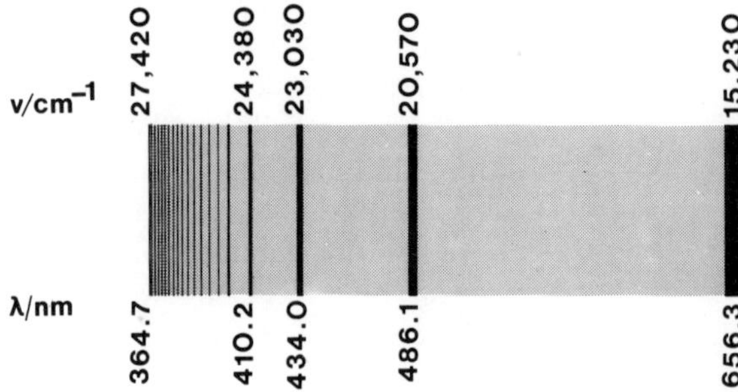

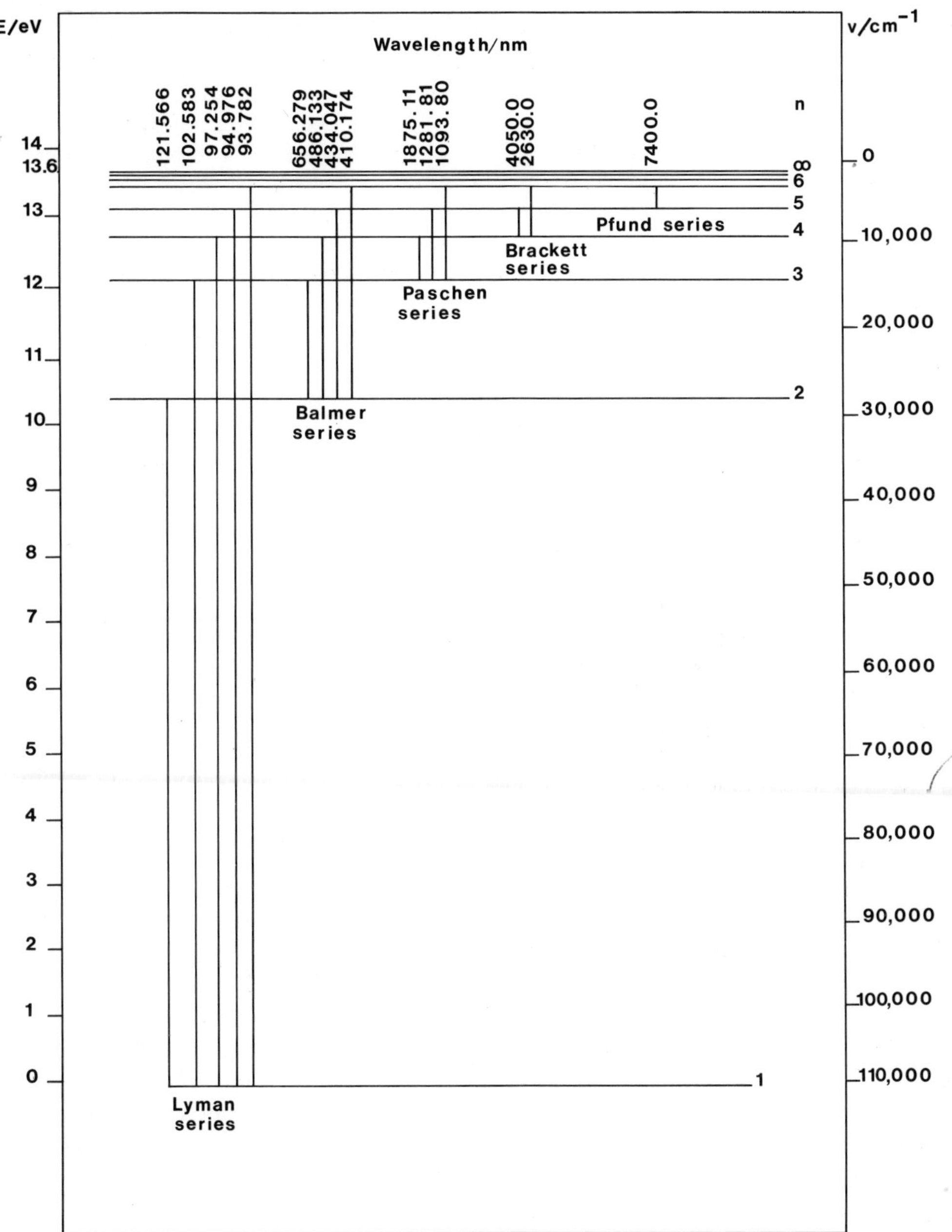

FIGURE 3.6. The hydrogen-atom energy-level diagram; only the most prominent lines are shown with wavelengths designated in nanometers (nm).

From the precision of Rydberg's constant it is obvious that the line spectrum of hydrogen atoms must be very sharp and can be measured with great accuracy.

The other series of lines in the hydrogen-atom spectrum could also be given by a simple formula similar to Equation 3.14.

$$\tilde{\nu} = R\left(\frac{1}{n_1^2} - \frac{1}{n_2^2}\right) \tag{3.16}$$

Here n_2 is again an integer, as is n_1. The values of n_1 increase monotonically from $n_1 = 1$ for the Lyman series to $n_1 = 5$ for the Pfund series.

We emphasize here the spectrum of the simplest atom, namely, hydrogen, points the way toward understanding more complicated atoms and molecules. This is so because hydrogen gives the simplest spectrum of all of the neutral elements, thus the interpretation should be the easiest. As a matter of fact, the spectra of many other elements could also be expressed by similar formulas such as

$$\tilde{\nu} = \frac{R}{(n_1 + d_1)^2} - \frac{R}{(n_2 + d_2)^2} \tag{3.17}$$

of which Equation 3.16 is, of course, just a special case. In Equation 3.17 d_1 and d_2 are additional constants specific for a given element, and n_1 and n_2 are again integers. For example, one series of spectral lines of the lithium atom has frequencies (in cm^{-1}) given by the formula

$$\tilde{\nu} = 28602 - \frac{109737}{(n + 0.59551)^2} \tag{3.18}$$

where n is an integer greater than 1.

Formulas of the type in Equations 3.14–3.18 were developed empirically by experimentalists. To develop a theoretical explanation of the spectra proved to be a formidable task. Historically the first major advance was made by Niels Bohr, a Danish physicist working in Rutherford's laboratory in 1913. The Bohr theory of atomic spectra and hence atomic structure *did not* stand the test of time and has long since been rejected in favor of the quantum-mechanical theory. Still, several ideas advanced as postulates were subsequently to show the way for quantum mechanics. It is worth our while to discuss briefly these concepts, particularly for the hydrogen atom, where Bohr's theory appeared at first to be strikingly successful.

The Bohr Theory of Atomic Structure

We mentioned previously that the Rutherford model of the atom presented theoretical difficulties because the classical laws of electrodynamics predict it to be unstable. We can show this quite simply in an analytical form. In Figure 3.7 our model can be shown as Rutherford and subsequently Bohr pictured it.

Let us consider what is needed to produce a dynamically stable atom. Firstly, to deflect the electron from straight-line motion and keep it moving in a circular path requires a centripetal force of:

$$F_{\text{cent}} = \frac{m_e v^2}{r} \tag{3.19}$$

Secondly, if the velocity and distance are suitable then the electrostatic force of attraction, which is

$$F_{\text{elec}} = \frac{e^2}{4\pi\epsilon_0 r^2} \tag{3.20}$$

will have just the required magnitude; that is, F_{elec} will be equal to F_{cent}. In this event we have:

$$\frac{m_e v^2}{r} = \frac{e^2}{4\pi\epsilon_0 r^2} \tag{3.21}$$

From this we can calculate the energy of the electron moving in the orbit. The total energy is the sum of kinetic, T, and potential, V, energy contributions:

$$E = T + V. \tag{3.22}$$

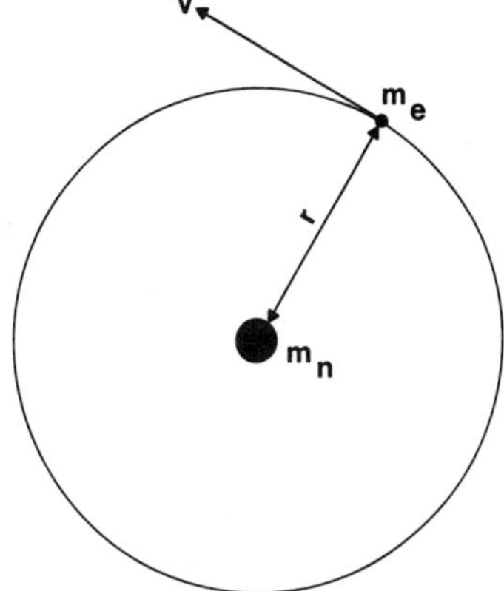

FIGURE 3.7.
Rutherford–Bohr model of the hydrogen atom. A single electron of mass m_e moving in a circular orbit with velocity v at a distance r from the nucleus of mass m_n.

The kinetic energy is that of motion

$$T = \frac{1}{2} m_e v^2 \tag{3.23}$$

whereas the potential energy is given by $F_{elec} = -dV/dr$, F_{elec} being given in Equation 3.20

$$V = -\frac{e^2}{4\pi\epsilon_0 r}. \tag{3.24}$$

Obviously from Equations 3.23 and 3.21 we obtain:

$$T = -\frac{V}{2} \tag{3.25}$$

so that the total energy is just

$$E = -\frac{V}{2} + V = -\frac{e^2}{8\pi\epsilon_0 r}. \tag{3.26}$$

Now electromagnetic theory says that an oscillating charge such as the rotating electron will radiate light and thus lose energy. From Equation 3.26, as the total energy of the electron steadily decreases, so does the radius r of the orbit of the electron. Thus the electron spirals into the nucleus, emitting continuous radiation of increasingly high frequency. This quite frankly just does not happen and hence does not provide an acceptable explanation of the structure of the hydrogen atom.

Enter Niels Bohr. He accepted the basic Rutherford model (see Figure 3.7) but rejected the classical interpretation just given. Instead he arbitrarily proposed that the orbits of the electrons in the hydrogen atom be *quantized*—that is, that only certain discrete orbits be allowed, and that when an electron is in a quantized orbit it does not emit radiation despite classical theory. He explained the spectrum of hydrogen by the electron jumping from one orbit to another emitting or absorbing a photon as it goes (according to the law of conservation of energy). This Bohr did by quantizing the angular momentum L, of the electron to integral multiples of $h/2\pi$, that is:

$$L = n\frac{h}{2\pi}. \tag{3.27}$$

Bohr used Planck's constant, which has the same dimensions of action[2] as does the angular momentum. The con-

[2]Action can either be expressed as energy multiplied by time (kg m^2 s^{-2} s) or as momentum by length (kg m s^{-1} m)

cept of angular momentum is perhaps new so we amplify it in Appendix I. As it turns out, the angular momentum of a particle of mass m moving in a circular path of radius r is given by

$$L = mvr. \tag{3.28}$$

Hence we have for an electron in a circular orbit

$$m_e vr = \frac{nh}{2\pi} = n\hbar \tag{3.29}$$

where n is an integer and we use the special symbol $\hbar$ (called *h-bar*) to represent $h/2\pi$ because this quantity occurs so frequently in quantum-mechanical descriptions.

The rest is simple. Solving for v in Equation 3.29 gives

$$v = \frac{nh}{2\pi m_e r} \tag{3.30}$$

and substituting into the expression for Newton's third law, Equation 3.21, we obtain:

$$\frac{m_e n^2 h^2}{4\pi^2 m_e^2 r^2} = \frac{e^2}{4\pi\epsilon_0 r} \tag{3.31}$$

or

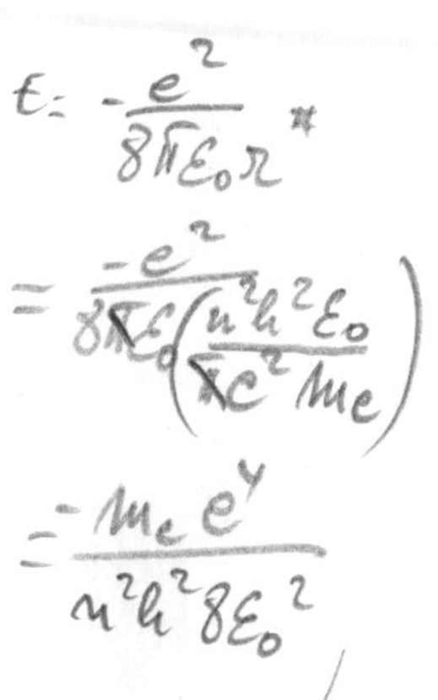

$$r = \frac{n^2 h^2 \epsilon_0}{\pi e^2 m_e} \tag{3.32}$$

giving the radius of the various quantized circular orbits allowed for each integral value of n. For the first orbit $n = 1$ we get:

$$\begin{aligned} r &= a_0 \\ &= \frac{(1)^2(6.6256 \times 10^{-34}\text{J s})^2(8.855 \times 10^{-12}\text{s}^2\text{C}^2\text{kg}^{-1}\text{m}^{-3})}{(3.1416)(1.602 \times 10^{-19}\text{C})^2(9.1091 \times 10^{-31}\text{kg})} \\ &= 5.293 \times 10^{-11}\ \text{m} = 52.9\ \text{pm}. \end{aligned}$$

The Bohr radius for $n = 1$ is often given the special symbol a_0.

Similarly, using the energy expression (Equation 3.26) with our quantization conditions gives for the hydrogen-atom energy:

$$E_H = -\frac{1}{n^2}\left(\frac{e^4 m_e}{8\epsilon_0^2 h^2}\right) \tag{3.33}$$

For $n = 1$ the energy of the electron of the first Bohr orbit is

$$E_H^1 = -2.179 \times 10^{-18}\ \text{J}.$$

Converting this to frequency units we find

$$E_{\mathrm{H}}^{1} = -3.290 \times 10^{15}\ \mathrm{Hz}$$

the value of R, Rydberg's constant, determined previously from spectroscopic experiments.

The integer n in the Bohr atom energy expression is known as a *quantum number.* We encounter other quantum numbers frequently and discuss them more fully later on. Briefly, quantum numbers specify discreteness, and this is needed to explain the line spectra of atoms.

For various values of the integer n in the Bohr energy expression (Equation 3.33) we get discrete energy levels, E_1, E_2, E_3, . . . , as shown in Figure 3.8. According to Bohr, only the states of the atom corresponding to these energies are stable. They are referred to as *stationary states* and Bohr assumed that, in contradiction to classical electromagnetic theory, no energy is radiated when an atom is in one of these stationary states if left to its own

FIGURE 3.8. Stationary states of the hydrogen atom calculated using the Bohr formula $E = -R/n^2$.

n	E/J	E/eV
6	-6.05×10^{-20}	−0.38
5	-8.71×10^{-19}	−0.54
4	-1.36×10^{-19}	−0.85
3	-2.41×10^{-19}	−1.51
2	-5.45×10^{-19}	−3.40
1	-2.18×10^{-18}	−13.60

devices. However, in the presence of electromagnetic radiation the atom can undergo a change in energy corresponding to a change from one stationary state to another. There is a gain in energy if light absorption occurs or energy loss if the hydrogen atom emits radiation. The frequency, ν, of the radiation absorbed or emitted is obtained from Planck's formula $E = h\nu$. Therefore, an atom absorbing energy radiation of frequency ν changes from its ith stationary state to its jth stationary state by

$$h\nu = E_j - E_i. \tag{3.34}$$

This expression is known as the *Bohr frequency condition*. A simple comparison of the differences $E_j - E_i$ of the Bohr stationary state energies in Figure 3.8 with the experimental hydrogen spectrum shown in Figures 3.5 and 3.6 shows that they correspond exactly. The integers n_1 and n_2 in the empirical formula of the spectroscopists (Equation 3.16) are merely Bohr's quantum number n for two different stationary states.

Stationary States Confirmed

Aside from spectra, a striking demonstration of the existence of discrete energy levels, or "stationary states," as Bohr called them, was given by the experiments of Franck and Hertz in 1914 on *critical potentials*. A schematic diagram of their experiment is shown in Figure 3.9. They filled the gap between the filament F and plate P with sodium vapor and bombarded the vapor with electrons of known energy. Until the energy of the electrons reaches a critical point the collisions with sodium atoms are elastic and increasing the accelerating potential, V_1, merely increases the the electron current flow at the collecting plate. However, when the accelerating potential reaches 2.10 V a sudden decrease in electron current is observed and the vapor begins to emit the characteristic yellow

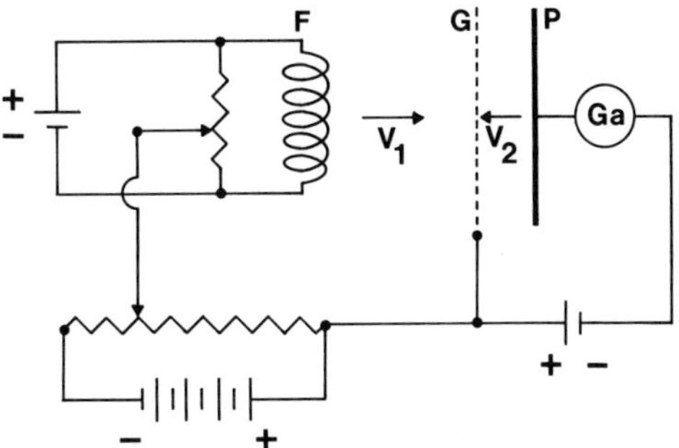

FIGURE 3.9.
Schematic diagram of Franck–Hertz experiment. F, filament providing the bombarding electrons; V_1, accelerating potential for electrons; G, grid; P, plate for collecting electrons (current indicated by galvanometer, Ga); V_2, retarding potential to prevent electrons that have undergone inelastic collison from reaching the plate.

color of the D-line of sodium (see Chapter 4 for an energy-level diagram of sodium). Now an energy of 2.10 eV corresponds to

$$(2.10 \text{ eV molecule}^{-1})(1.60 \times 10^{-19} \text{ J eV}^{-1}) = 3.36 \times 10^{-19} \text{ J molecule}^{-1}$$

or

$$= 1.69 \times 10^{4} \text{ cm}^{-1}$$

while the D-line of sodium is 589.3 nm = 1.697×10^4 cm^{-1}. Similar results were found for other atomic vapors.

What happens here is that when the energy of the bombarding electrons corresponds to the energy difference between stationary states, inelastic collisions occur. The energy of the collision is used to excite the atom to the upper stationary state and the electrons are not left with enough kinetic energy to reach the collector plate. This potential, at which a fall in current is detected by the galvanometer, G, is termed a *critical potential.*

The Demise of the Bohr Theory

Although it worked so nicely for the simple hydrogen atom spectrum, the Bohr theory met its Waterloo on three main accounts:

1. The *assumption* of stationary states without any theoretical justification provides no basis for real understanding of the quantization process that occurs.
2. The Bohr theory could not predict the spectra of any other element, such as helium or other more complex atoms, let alone any molecules.
3. When a magnetic field is present, the hydrogen-atom spectrum shows splittings of some lines and not of others. The Bohr theory predicts that *all* lines will split because the energy states all have angular momentum.

The development of the new quantum mechanics showed many other defects in the Bohr theory, as we see shortly. Still, Bohr deserves just credit, because his stationary state concept, together with de Broglie's hypothesis concerning the wave nature of matter, $\lambda = h/p$, enabled a German scientist, E. Schroedinger, to come up with a satisfactory formulation of quantization in 1926.

Schroedinger answered the riddle of atomic structure by recognizing that there already existed a branch of physics in which integral numbers arise quite naturally, namely, the theory of wave motion as it applies to standing waves. It is worth our while, before looking at Schroedinger's wave mechanics, which became known as "quantum mechanics," to examine a couple of physical systems where standing waves exist. We choose the vibrating string exemplified by a guitar string and the vibrating membrane illustrated by a drum head.

The Standing Waves of a Vibrating Guitar String

A vibrating string of a guitar represents a physical system where only a discrete set of states of vibration are possible, and not any arbitrary vibrational motion.

The guitar string is fixed by the bridge at one end and the fingers on the frets at the other end. By plucking the string the player sets up acoustical vibrations of certain characteristic frequencies that are determined by the length of the string. These *audio*frequencies determine the pitch and quality of the guitar. That only certain vibrations are allowed is a result of the string being fixed at the ends. This condition is known as a *boundary condition*. The vibrations can be represented as sine waves in the manner shown in Figure 3.10. They have zero velocity and are known as standing waves. The boundary condition in this case is that the standing wave *must* have zero amplitude at the two ends—it is fixed at these points and cannot move.

The simplest vibration in Figure 3.10(a) is called the *fundamental* vibration and determines the pitch or note of the string. In the next simplest vibration [Figure 3.10(b)] the string has one complete wavelength λ. Here, as the string vibrates, each point passes through zero amplitude. Points along the string that do not vibrate, i.e., have constant zero amplitude, are known as *nodes*. The points of maximum amplitude at any given time are called *antinodes*. This vibration, together with those of increasing number of nodes [Figure 3.10(c), (d)], give the guitar its quality or "timbre" and are known as the *harmonics of the fundamental vibration*. Proper blending of the fundamental vibration with its harmonics is what makes a good guitar.

If the length of the string is L, and the distance between nodes is obviously $\lambda/2$, we must have the relationship

$$L = n\,\frac{\lambda}{2} \tag{3.35}$$

for the guitar string. When $n = 1$ we get the fundamental vibration, $n = 2$, the first harmonic, and so forth. The number n must be an integer in the same way it was an integer in Bohr's atomic theory, and if we like we can think of it as a quantum number for the guitar string. There is no way that n can be anything but integral because nonintegral values of n lead to destructive interference of the vibrational waves.

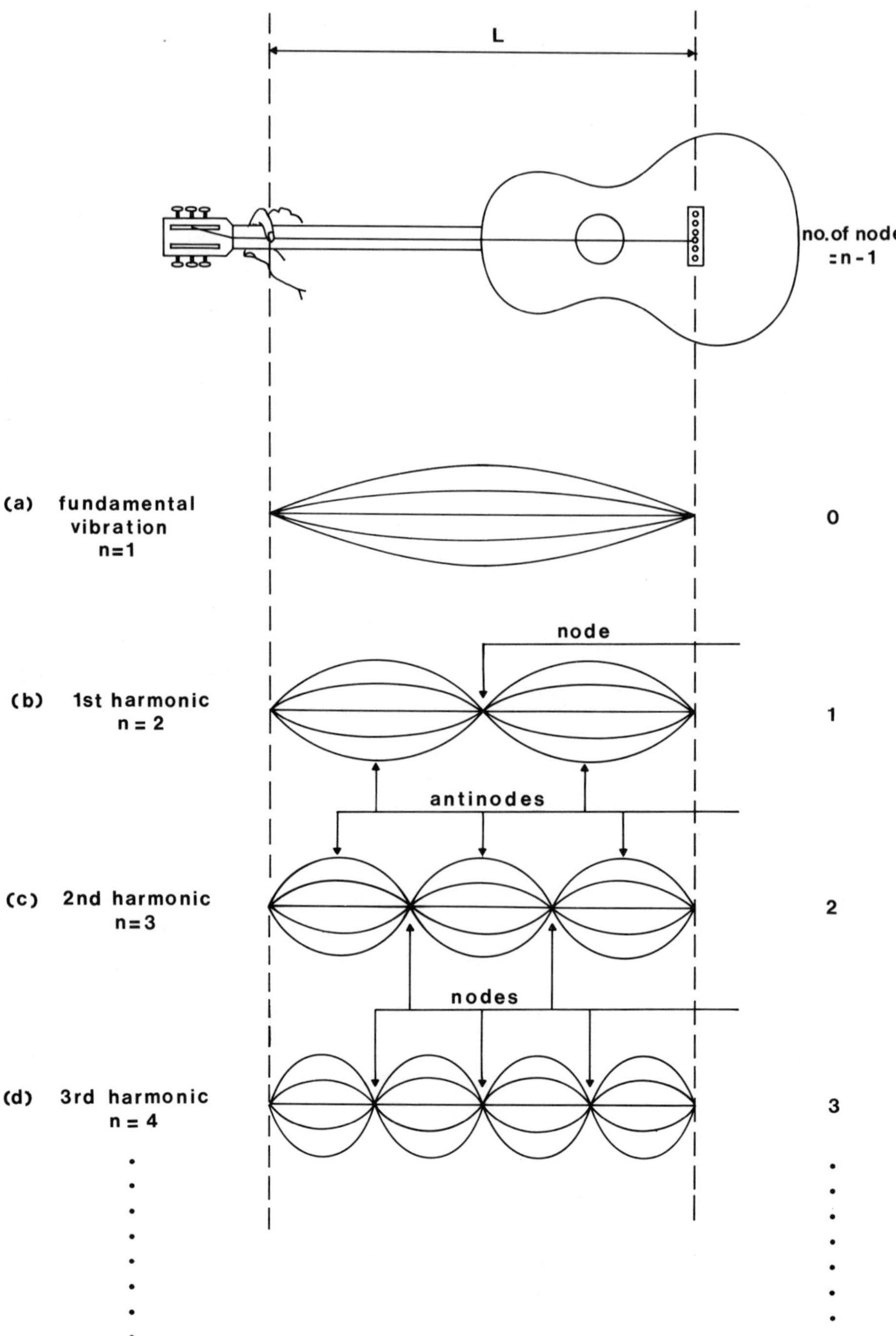

FIGURE 3.10. The vibrations of a guitar string.

The Standing Waves of a Vibrating Drum Head

A drum head can be treated in an analogous manner to the guitar string, except that we now have a two-dimensional system rather than one-dimensional.

The boundary conditions for the vibrational or acoustic waves of a drum are that the amplitude must be zero around its periphery. Now suppose that the drum is struck in the center, as shown in Figure 3.11. Again there is a fundamental vibration and various harmonics that accompany it.

The vibrations in this case are called *radial* vibrations and the circular symmetry of the drum head is maintained at all times. Notice that as with the guitar string there is a change of phase (sign) across each nodal line. The nodes in this case are called *radial nodes*.

A different set of vibrations is set up if the drum membrane is struck off-center. These vibrations do not retain the circular symmetry of the membrane but extend across the drum, as shown in Figure 3.12, and are known as *angular* vibrations. Here the nodes shown are known as *angular nodes*.

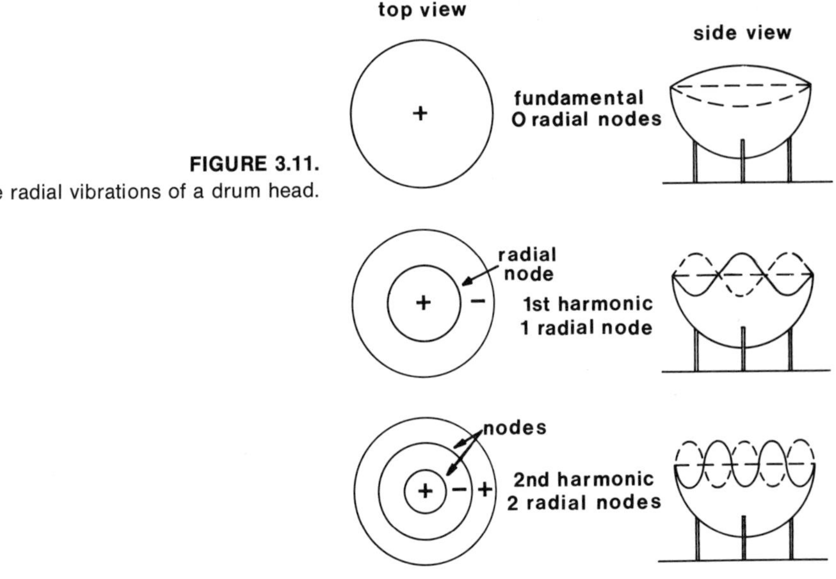

FIGURE 3.11.
The radial vibrations of a drum head.

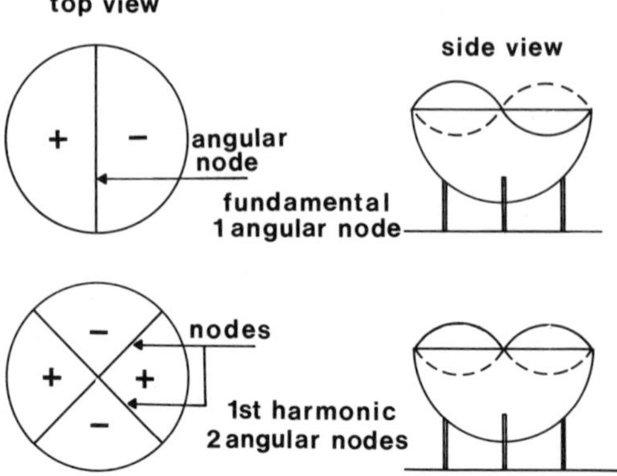

Figure 3.12.
The angular vibrations of a drum head.

The quantization of the drum-head vibrations is considerably more complicated mathematically than that for a simple string but again, without going into detail, the equation of the drum-head vibrations involves two integers, because the drum vibrations are two-dimensional. One is specific to both the angular and radial modes of vibration, and the other is for just the angular modes. The point we wish to make, however, still holds. Only certain discrete vibrations are allowed for the drum, just as only discrete vibrations are allowed for a guitar string.

A Wavelike Equation Applied to Matter

Schroedinger recognized that by formulating an equation of motion for atoms and molecules that incorporated the wave properties of matter, one *automatically* introduces the quantum numbers in the same way that they arise for guitars and drums. He correctly formulated the so-called wave mechanics of matter.

An equation of motion involves the principle of conservation of energy. Classically, we have seen this is simply that the total energy of a system, that is, the sum of its kinetic and potential energies, is constant.

$$E_{\text{tot}} = T + V = \text{const.} \tag{3.36}$$

Substituting $T = (mv^2/2)$ and leaving the potential energy unspecified, we get:

$$E_{\text{tot}} = \frac{mv^2}{2} + V \tag{3.37}$$

or since the momentum $p = mv$:

$$E_{\text{tot}} = \frac{p^2}{2m} + V. \tag{3.38}$$

This equation was transformed into a wave equation by Schroedinger as follows: wherever p appears we replace it by a differentiation, $-i\hbar(d/dx)$, $\hbar$ being Planck's constant divided by 2π and $i = \sqrt{-1}$. In other words, symbolically:

$$p \rightarrow -i\hbar \frac{d}{dx} \tag{3.39}$$

Then we have an equation that reads:

$$\frac{1}{2m}\left(-\hbar^2 \frac{d^2}{dx^2}\right) + V = E. \tag{3.40}$$

Of course, Equation 3.40 is meaningless as it stands, because it only tells us to differentiate twice with respect to x. We have to know *what* to differentiate. Let us call the thing we operate on a *wave function* and give it the symbol ψ. Equation 3.40 will now read

$$-\frac{\hbar^2}{2m}\frac{d^2\psi}{dx^2} + V\psi = E\psi. \tag{3.41}$$

This rather complicated second-order differential equation is known as the *Schroedinger equation*. It is very similar to ordinary wave equations of motions that one might write for the motions of a guitar string or a drum head except that it incorporates Planck's constant h. The solution in the case of the guitar string gives a wave function that describes the standing waves of the string as:

$$\phi = A \sin \frac{n\pi x}{\lambda} \tag{3.42}$$

A being the maximum amplitude of the wave motion, λ the wavelength, and ϕ the amplitude of the motion at a given point along the string, x. For atoms our wave function ψ in Equation 3.41 contains our knowledge of the electron in a one-electron atom in one dimension only, and to generalize to three dimensions we would have to include the y and z coordinates as well. For our purposes, however, the Schroedinger equation even in one dimension is mathematically complex, as are its solutions, so that from here on we discuss the results in a qualitative fashion only.

Solutions to Schroedinger's Wave Equation

There are two quantities of interest that we obtain by solving the second-order differential equation such as Equation 3.41. Firstly, there are discrete values of the energy, E, of the system. They are discrete because, just as in the guitar string and drum membrane, there are boundary conditions that we impose on the equation. These boundary conditions are very similar in this case and are as follows:

1. The value of ψ at infinity is zero. This is the analog of ψ vanishing at $x = 0$ and $x = L$, the string length, in Equation 3.42.
2. The value of ψ must be continuous and vary in a smooth fashion, not jumping about. This is similar to the requirement for a vibrating string that the string not be broken!
3. The value of ψ must be such that in the case of an electron, for example, the probability of finding it somewhere in all space is finite.

Secondly, we get values for the wave function itself, ψ. Because ψ is describing wavelike motion it is going to involve trigonometric sine or cosine functions and/or exponential functions. We expect our quantum-mechanical wave function to have similar functional dependence to that of Equation 3.42, but again to probably involve the quantum constant of Planck h.

In a guitar string the wave function ϕ gives us the amplitude of the vibration, but the square of this amplitude gives us the more physically meaningful quantity, the intensity of the sound. So too, by analogy, it is the square of the quantum-mechanical wave function ψ^2 in the Schroedinger equation that has physical significance.[3] The quantum mechanical wavefunction itself, ψ, is nothing more than a mathematical function.

In terms of electrons the interpretation of the wave function ψ is this:

> The probability of finding the electron within the volume range of v to $v + dv$ is given by
>
> $$|\psi|^2 \, dv.$$

Thus when $|\psi|^2 \, dv$ is a large number, the probability of finding the electron in that volume element is high. When $|\psi|^2 \, dv$ is small, the probability is low. At nodal points or

[3]Sometimes the analytical form of ψ is complex, involving $\sqrt{-1}$. To make it always positive we use the square of the modulus, $|\psi|^2$.

lines where ψ is always zero $|\psi|^2\,dv$ will be vanishingly small.

To put this idea across more clearly imagine a gigantic blow-up of a hydrogen atom, so huge that our infinitesimally small volume element becomes as large as a matchbox. Around the nucleus let us pack together a large number of matchboxes so that there is no empty space left. The experimenter now opens a particular matchbox to see if the electron is inside at a given instant in time. The probability of finding the electron in any given matchbox is $|\psi|^2$. The value of $|\psi|^2$ will vary with each matchbox, depending on whether the electron is more or less likely to be found therein.

There are several ways to represent probability densities such as $|\psi|^2$ in a pictorial manner that help to clarify the physical reality of this function. As an example we take the simplest wave function of the hydrogen atom, termed the 1s wave function.

Its mathematical form is:

$$\psi_{1s} = \frac{1}{\sqrt{\pi}a_0^{3/2}} \exp\left(\frac{-r}{a_0}\right) \tag{3.43}$$

where a_0 is $= 52.9$ pm and r represents the distance of the electron from the nucleus. Squaring this function gives:

$$\psi_{1s}^2 = \frac{1}{\pi a_0^3} \exp\left(\frac{-2r}{a_0}\right). \tag{3.44}$$

We can then plot ψ_{1s}^2 against r to obtain, as in Figure 3.13, the probability density of the electron at any distance r from the nucleus. This curve shows that the maximum probability of finding the electron occurs at the nucleus itself and then becomes less and less as the distance from the nucleus becomes larger and larger.

Another way of representing the electron's probability density given by ψ^2 is to use stippling as in Figure 3.14, where each dot represents a hypothetical instantaneous picture of the electron's position, then an increasing number of pictures being superimposed to give the resulting "dartboard" picture. Obviously the electron does not have a definite radius from the nuclei such as in the Bohr theory. Rather it can only be characterized as a fuzzy "electron cloud" or more correctly, as a probability function.

It may seem strange that the most probable point of finding the electron is at the nucleus itself for this particular hydrogen wave function. However, the student is

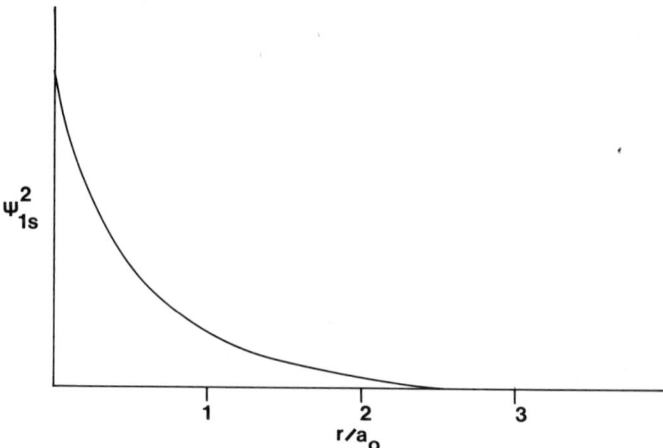

FIGURE 3.13. Plot of $\psi^2 = (1/\pi a_o^3)\exp(-2r/a_o)$ versus r/a_o giving the probability of finding an electron at any distance r from the nucleus.

reminded that we must talk about *probability per volume element* where each volume element has a finite volume. To illustrate this we plot $4\pi r^2\,\psi_{1s}^2$ against r in Figure 3.15. Here we are representing graphically the probability of finding the electron in concentric shells of width dr and

FIGURE 3.14. A stippling representation of ψ_{1s}^2 showing the probability of finding the electron around the positive nucleus.

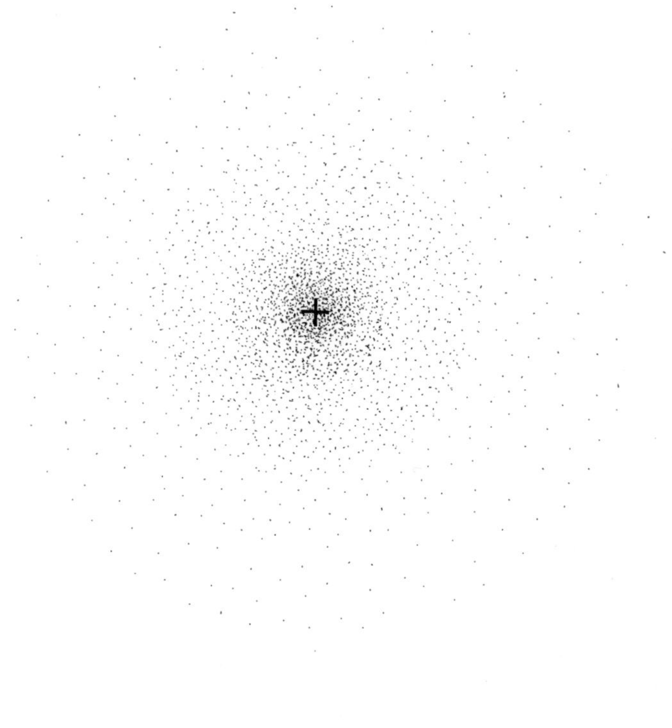

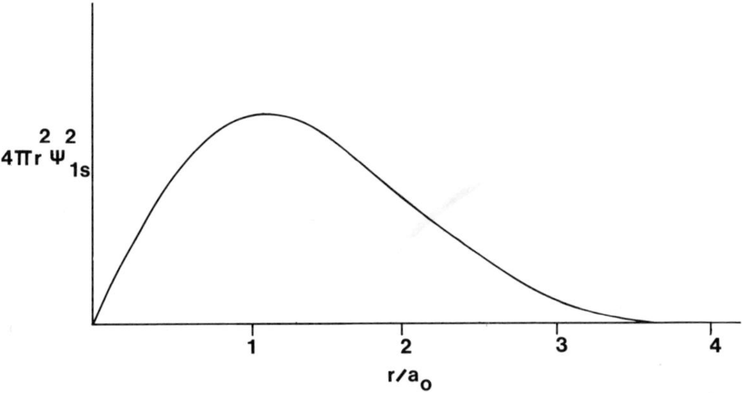

FIGURE 3.15. Plot of $4\pi r^2\ \psi_{1s}^2$ versus r/a_o giving the probability distribution.

volume $4\pi r^2\ dr$ as r increases. This plot is called a *probability distribution* and it shows that the probability of finding the electron in the volume elements near the nucleus become vanishingly small because the volume elements also become vanishingly small.

The Heisenberg Uncertainty Principle

One of the consequences of quantum mechanics is that it places certain limitations on our knowledge of the behavior of electrons and other atomic-sized entities. This arises as a direct result of the dual nature of matter and was first recognized by W. Heisenberg in 1927. He first stated what is now known as the *Heisenberg uncertainty principle*.

> For an electron there is a fundamental limitation to the precision with which one can measure simultaneously both its position and momentum.

The principle applies to all matter in reality, but only takes on significance for particles in the quantum realm. Heisenberg showed that the lower limit for our observational knowledge for position and momentum is the Planck constant h divided by 4π. We can express this as:

$$\Delta p\ \Delta q \geqslant \frac{\hbar}{2} \tag{3.45}$$

Where Δp represents the uncertainty in momentum and Δq the uncertainty in position.

We can see to some extent just why this principle applies by remembering the particulate nature of light and the momentum of the photon, h/λ (see the de Broglie relation, Equation 3.13). We use light to give us an image of any object that we are trying to see. The precision with which such an observation of position can be made is limited to one wavelength of the light, $\Delta q \approx \lambda$. If we represent the light photon as a particle, it hits the electron when trying to measure its position but simultaneously changes the momentum of both the probing light photon and the electron. The electron's momentum will be altered by an amount similar to the original momentum of the photon, $\Delta p \sim h/\lambda$. Thus we can see that the product of Δq and Δp is roughly $\lambda(h/\lambda) = h$. It requires more detailed mathematical analysis to obtain the exact Equation 3.45, but the qualitative discussion given here shows that the right-hand side should be roughly of the magnitude of Planck's constant. It also shows that the shorter the wavelength one uses in an attempt to improve the precision of the position measurement, the more one will alter the momentum of the electron during the position measurement and vice versa. This is illustrated in Figure 3.16.

That the uncertainty principle is a strictly quantum principle can be illustrated by considering a cricket ball weighing, say, 0.2 kg and traveling at a speed of 5.0 m s^{-1}, thus having a momentum of 1.0 kg m s^{-1}. Imagine now that we can measure the momentum using stroboscopic lights to a precision of one part in 10^{12}; hence the uncertainty in momentum would be 10^{-12} kg m s^{-1}. Thus the uncertainty in position is then according to the uncertainty principle:

$$\begin{aligned}\Delta q &= \frac{6.6 \times 10^{-34}\ \text{kg m}^2\text{s}^{-1}}{(4)(3.14)(10^{-12}\ \text{kg m s}^{-1})}\\ &= 5 \times 10^{-23}\ \text{m}\end{aligned}$$

FIGURE 3.16. The Heisenberg uncertainty principle. The very short wavelength photon with momentum m_1v_1 attempts to locate the position of the electron with momentum m_ev_e. Simultaneous change in the electron's momentum to m'_ev_e' renders precise knowledge of position and momentum impossible.

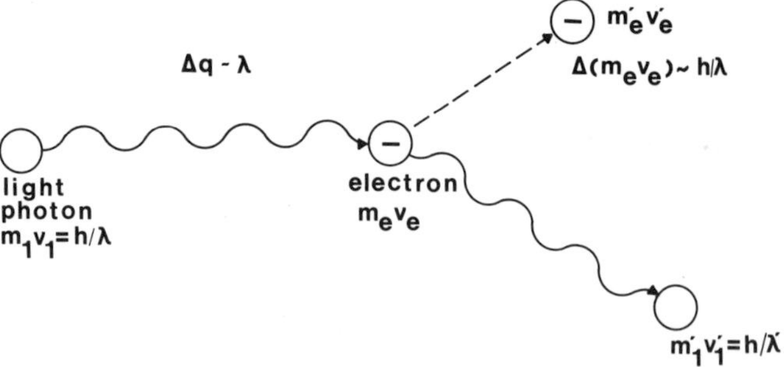

which is completely negligible (far smaller than the diameter of an atomic nucleus!).

The picture for an electron is quite different. Say that we hoped to study in detail the motion of the electron in a hydrogen atom and wanted to ascertain its position within about $\frac{1}{10}$ of a Bohr radius, that is, to within 5 pm. Then the uncertainty in momentum becomes

$$\Delta p = m\,\Delta v = \frac{6.6 \times 10^{-34}\ \text{kg m}^2\text{s}^{-1}}{(4)(3.14)(5 \times 10^{-12}\text{m})}$$
$$= 1.05 \times 10^{-23}\ \text{kg m s}^{-1}$$

so
$$\Delta v = \frac{1.05 \times 10^{-23}\ \text{kg m s}^{-1}}{9.01 \times 10^{-31}\ \text{kg}} = 1 \times 10^{7}\ \text{m s}^{-1}$$

which is a huge uncertainty in velocity—nearly $\frac{1}{30}$ the speed of light!

Obviously our interpretation of the quantum-mechanical wave function in terms of probability is integrally related to the Heisenberg uncertainty principle. We cannot state in atoms or molecules the exact location of an electron at a given instant in time. The best we can do is to conjecture that it most probably will be in some volume element at a given time. We cannot talk about trajectories of the electron around the nucleus as we would if it were a Bohr-type atom. In wave mechanics, only probabilities count and we must readjust our classical way of observing and imagining objects on an atomic scale and think of them in this quantum-mechanical way.

Quantum-mechanical Hydrogen Atom

The Schroedinger equation for the hydrogen atom describes the motion of the single electron in three dimensions. Since our one-dimensional guitar string used one quantum number and the two-dimensional drum head used two quantum numbers, it is not surprising that our quantum-mechanical description of a three-dimensional atom will use three quantum numbers to describe the electron's motion. They are denoted by the symbols, n, l, and m_l and have the following significance:

n—called the *principal quantum number.* It can take any integral value 1,2,3,4 . . . and determines the energy of the hydrogen atom according to the formula

$$E_n = -\frac{RZ^2}{n^2} \tag{3.46}$$

R being Rydberg's constant $=$ 109,737 cm^{-1}, Z the nuclear charge ($= +1$ for hydrogen), and n the principal quantum number = 1,2,3,4. . . . You will notice that Equation 3.46 is exactly the same energy expression as we found in Bohr's theory, which is satisfying because, after all, Bohr's theory did correctly predict the hydrogen-atom energies.

l—called the *orbital angular momentum quantum number.* It can take any integral values including zero up to $n - 1$ for a given value of n. For example, if $n = 3$, l can have values 0, 1, and 2. The shape of the region in which the electron moves is determined by n and l. The radial extension is mainly determined by n. The angular distribution is determined by l. Sometimes this region is spherical, as we have seen: at other times it is dumbbell-shaped, and so on.

m_l—called the *magnetic quantum number.* It actually represents a component of orbital angular momentum along a specified direction (usually, by convention, the z-direction). It can take any integral value from $-l$ to $+l$. For example, if $l = 2$, m_l can have values -2, -1, 0, $+1$, and $+2$. Thus the quantum number, m_l, has $2l + 1$ values. It determines the directional properties of the region in which the electron moves, for example, whether it moves primarily near the x, y, or z axis.

These three quantum numbers, n, l, and m_l give us all kinds of information about how and where the electron moves subject to the restriction of the Heisenberg uncertainty principle. Valency experts like to draw pictures that represent the electron's behavior for various values of the quantum numbers and we often loosely use the term *orbitals* for these pictorial representations. The word orbital is obviously a throwback to the old Bohr picture of orbits and much confusion exists in textbooks about the appropriate definition of an orbital. We take the definition of an orbital as follows:

> *orbital* (ôr′bĭt el) *n.* The various wave functions, ψ, for the hydrogen atom. Often represented pictorially as a contour[4] of constant $|\psi|$ such that the total charge outside the contour is small (e.g., 10%). Each orbital is specified by the three quantum numbers n, l, and m_l.

[4]A contour diagram is a means of representing a function of several variables by drawing lines on which the function has a constant value. Weather maps are a well-known example where lines of constant value of the atmospheric pressure are drawn on the map. Contour diagrams representing orbitals usually include the contour line for just one fixed value of $|\psi|$, so that regions of positive values of ψ and of negative values of ψ are both indicated.

The confusion arises because often orbitals are represented as ψ^2 rather than ψ, as we have defined them above. Certainly, as we have said previously, it is only ψ^2 that has any direct physical significance. Nonetheless, in subsequent chapters dealing with valency principles it is ψ, not ψ^2, that must be used. We draw pictures only to show the various properties of these orbitals.

So the functions ψ for the hydrogen atom are called *orbitals*. What do they look like and what are the relationships of the orbitals to the quantum numbers?

The 1s Orbital

We begin with the simplest orbital, for which $n = 1$, $l = 0$, and $m_l = 0$. It is called either ψ_{1s} or simply the ls orbital. The mathematical expression for this orbital was given earlier in Equation 3.43. Our orbital representation can have several forms, three of which are shown in Figure 3.17. Figure 3.17(a) is easily the most common way of pictorially drawing the ls orbital. It is a plane cut through a contour diagram of some constant value of ψ. If we imagine the orbital using spherical polar coordinates (see Appendix II) it has zero angular dependence, that is, it is spherically symmetrical. Figure 3.17(b) plots ψ_{1s} against r/a_0; in other words, the value of ψ as we go out from the nucleus along a line ($r = 0 \rightarrow r = \infty$). The values of ψ are given by the same plot whatever line we choose; that is, the orbital is spherically symmetrical. This sort of plot we refer to as a *radial dependence plot*. The plot in Figure 3.17(c) is similar except that here $4\pi r^2\psi_{1s}^2$ is plotted. Figure

FIGURE 3.17. A hydrogen 1s orbital. (a) Contour diagram showing zero angular dependence; (b) radial dependence; (c) radial dependence probability distribution.

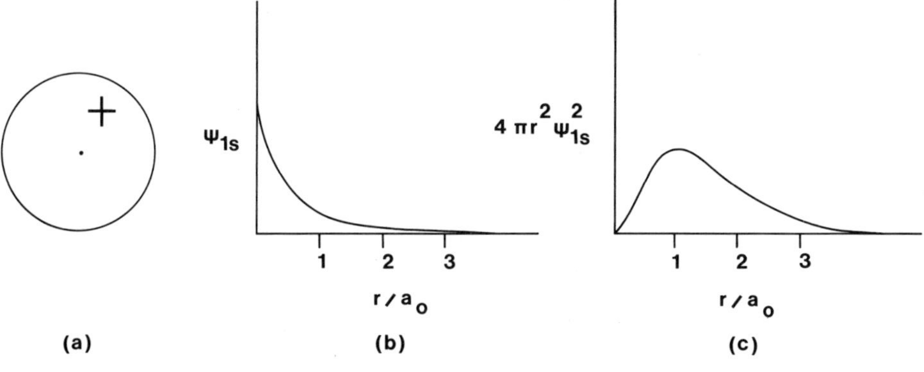

3.17(c) is called a *radial dependence probability* distribution plot and shows that the most probable distance of the electron from the nucleus is $r/a_0 = 1$, that is, $r = 52.9$ pm, in accord with the Bohr theory.

The notation "1s" is a shorthand for the quantum numbers involved in this orbital. The number 1 stands for the value of the principal quantum number n. The symbol s stands for the value of l according to

value of l	0,1,2,3,4 . . .
symbol	s,p,d,f,g . . .

We use only the first four of these symbols, which came from the adjectives *sharp, principal, diffuse,* and *fundamental* used by early spectroscopists to identify series of spectral lines before the birth of quantum theory. These adjectives were subsequently found to be connected with orbitals having l values of 0, 1, 2, and 3 respectively.

Quantum mechanics tells us that the average distance $\bar{r}$ of the electron from the nucleus is given by the expression:

$$\bar{r} = \frac{a_0 n^2}{Z}\left[\frac{3}{2} - \frac{l(l+1)}{2n^2}\right] \tag{3.47}$$

and in the case of the 1s orbital, this is simply

$$\bar{r} = \frac{3}{2}a_0 = \frac{3}{2}(52.9)\text{ pm} = 79\text{ pm}.$$

Notice that $\bar{r}$ is the average distance, not the radius itself, because with our proper quantum-mechanical formulation we can only speak of the whereabouts of the electron in terms of probabilities.

The 2s and 2p Orbitals

We move on to the next simplest case, namely, that when $n = 2$. Now both l and m_l can have values other than zero. Hence we end up with not just one orbital, but four. Specifically these are given in Table 3.1. The designation now includes the m_l quantum number as a subscript,

Table 3.1 Orbitals for $n = 2$

n	l	m_l	Orbital	Real orbitals
2	0	0	2s	2s
2	1	+1	$2p_1$ }	$2p_x$
2	1	−1	$2p_{-1}$ }	$2p_y$
2	1	0	$2p_0$	$2p_z$

which when $n = 2$ and $l = 1$ can be nonzero. The orbitals $2p_1$, $2p_0$, and $2p_{-1}$ obviously are directed differently in space because this is the physical meaning of the m_l quantum number. It turns out that the mathematical expressions for $\psi_{2p_{+1}}$ and $\psi_{2p_{-1}}$ involve imaginary numbers and, therefore, the spatial properties of these orbitals are difficult to visualize. However, the ψ_{2p_0} orbital is real and is directed along the z-direction in a Cartesian coordinate system. It is more generally known as the $2p_z$ orbital. By an appropriate transformation we can also construct two real orbitals from the $2p_{+1}$ and $2p_{-1}$ orbitals. Not surprisingly, these are directed along the x and y coordinates in a manner similar to the $2p_z$ orbital being directed along the z direction. This transformation of imaginary to real orbitals is further amplified in Appendix III. Figure 3.18(a) shows a cross section of the double sphere-like angular dependence of the three 2p orbitals, Figure 3.18(b) shows the

FIGURE 3.18. (a) The three 2p ($n = 2$, $l = 1$) orbitals for the hydrogen-atom angular dependence; (b) radial dependence; (c) radial dependence probability distribution.

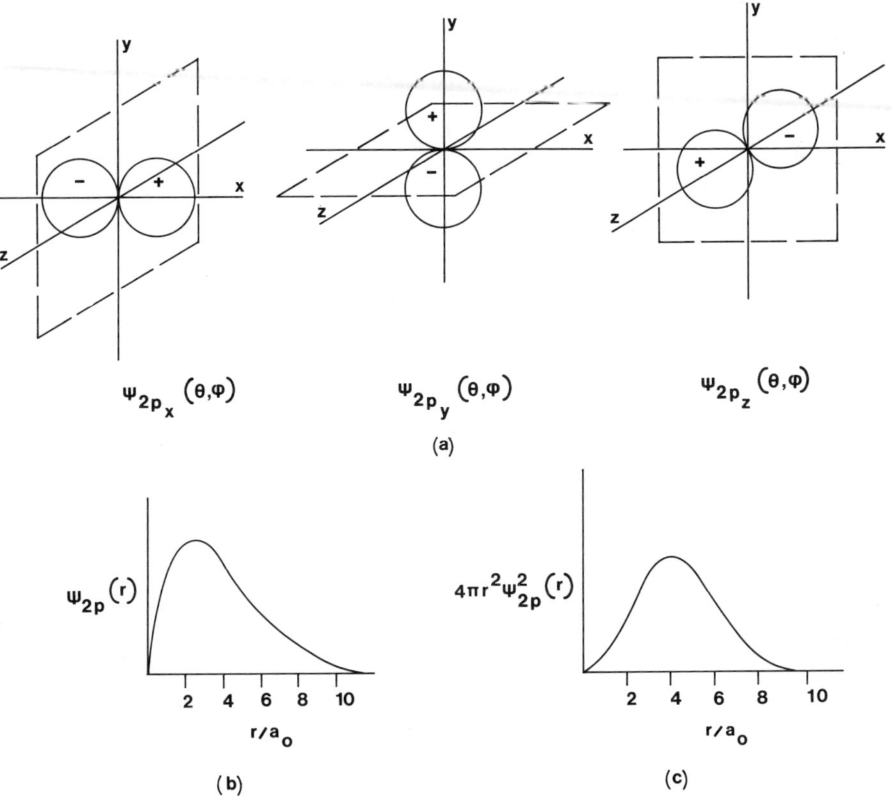

radial dependence, and Figure 3.18(c) shows the radial dependence probability distribution.

We can also include *both* the angular and radial dependence in our pictorial representation and so have the 2p orbitals as a contour of constant $|\psi|$ as shown in Figure 3.19. Here we illustrate a 2p orbital as a three-dimensional contour.

The diagrams showing just the angular dependence or those showing the contour surface of constant ψ are the ones we commonly use as our pictorial representation of the 2p orbitals. They are dumbbell-shaped and have a nodal plane going through the nucleus. This is an angular node and it shows that the wavefunction changes sign as designated by the plus and minus signs on the ''lobes'' of the orbital. We must be careful *not* to associate these algebraic signs with charge. What the signs mean is that the orbital, that is, the function ψ, has a positive value in some places and a negative value in others (just as $\sin\theta$ has a positive value for $0 < \theta < \pi$ and a negative value for $\pi < \theta < 2\pi$).

The remaining $n = 2$ orbital is an s type since $l = 0$. It is logically termed the *2s orbital,* and the orbital representations are shown in Figure 3.20. Again there is a node where the orbital changes sign. It is a radial node very analogous to the radial nodes of a drumhead. The number of radial nodes for any orbital is given by $n-l-1$.

A simple calculation with Equation 3.45 shows that the energy of all these four orbitals is the same, namely, $-\frac{1}{4}R$, R again being Rydberg's constant. When different orbitals have the same energy they are said to be *degenerate*. In the case of the four $n = 2$ orbitals above, the degeneracy is said to be fourfold.

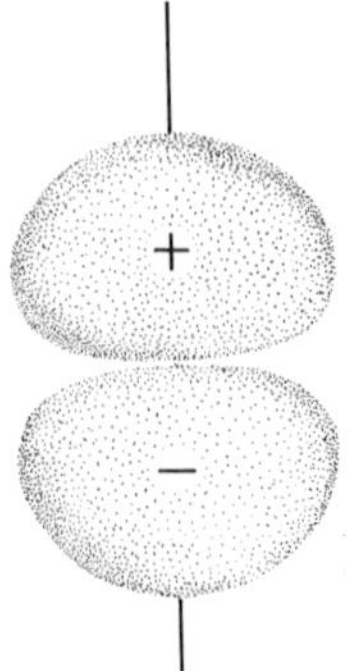

FIGURE 3.19.
A three-dimensional plot of a 2p orbital showing a contour of constant ψ.

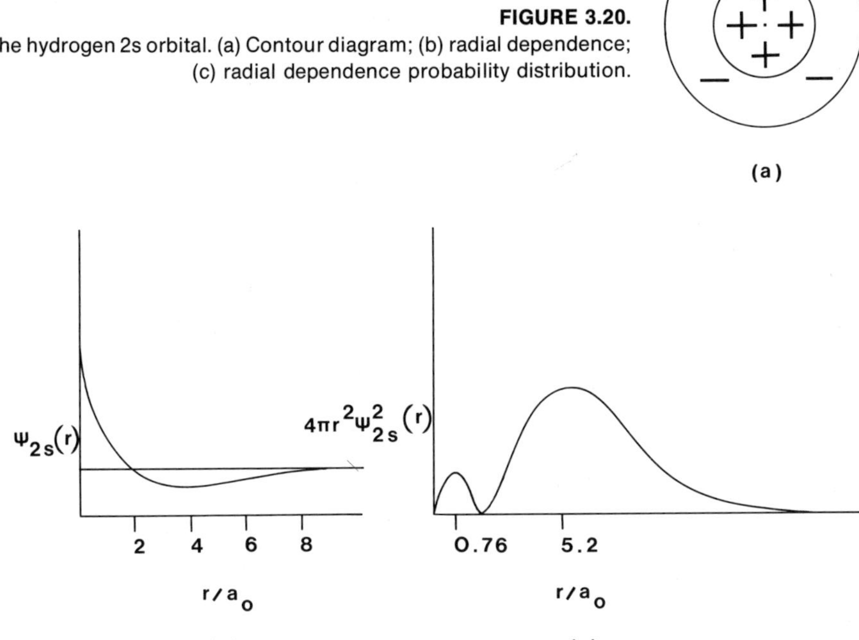

FIGURE 3.20.
The hydrogen 2s orbital. (a) Contour diagram; (b) radial dependence; (c) radial dependence probability distribution.

To get some idea as to the size of the orbitals with $n = 2$ we can again calculate the average distance $\bar{r}$ of the electron from the proton in a hydrogen atom using Equation 3.47.

Orbital	E	$\bar{r}$
2s	$-\frac{1}{4}R$	$6a_0$
$2p_x$	$-\frac{1}{4}R$	$5a_0$
$2p_y$	$-\frac{1}{4}R$	$5a_0$
$2p_z$	$-\frac{1}{4}R$	$5a_0$

Notice that the orbitals with $n = 2$ are all more diffuse than the 1s orbital in the sense that the average distance, $\bar{r}$, of the electron from the nucleus is up to four times as great. This is quite general; orbitals enlarge as n increases (provided that Z, the nuclear charge, does not change). Notice also that $\bar{r}$ is smaller for the 2p orbitals than for the 2s orbital.

Other relationships involving the principal quantum number n are also emerging.

1. The total number of nodes in any orbital is given by $n - 1$. Thus a 1s orbital has zero nodes, but a 2s has one node, as do the 2p orbitals.
2. The number of degenerate orbitals for the hydrogen atom for each value of n is n^2. Thus the degeneracy for $n = 2$ is fourfold, and we expect there to be a ninefold degeneracy for $n = 3$.

The 3s, 3p, and 3d Orbitals

The arguments presented above extend quite nicely to cover the orbitals when $n = 3$. They are tabulated in Table 3.2. There are nine degenerate orbitals, exactly as we predicted, and they are even more diffuse than the $n = 2$ orbitals.

Again the $3d_{+2}$, $3d_{+1}$, $3d_{-1}$, and $3d_{-2}$ orbitals involve imaginary numbers and we make the orbitals real as we did for the p-orbitals in Appendix III. The real d-orbitals are known as $3d_{z^2}$, $3d_{xz}$, $3d_{yz}$, $3d_{xy}$, and $3d_{x^2-y^2}$ and are represented in our pictorial fashion in Figure 3.21 along with the 3s and 3p orbitals.

n = 4 Orbitals

Carrying on to the 16-fold degenerate orbitals for the hydrogen atom that occur when $n = 4$, we see that these consist of one 4s, three 4p, five 4d, and seven 4f orbitals. The s, p, and d orbitals are similar to those that we have already described except they all have one additional radial node. The 4f orbitals are more difficult to draw but it can be done [see, for example, *J. Chem. Educ.* (1964) *41*,

Table 3.2 Orbitals for $n = 3$

n	l	m_l	Orbital	Real orbital	E	$\bar{r}$
3	0	0	3s	3s	$-1/9R$	13.5 a_0
3	1	1	$3p_1$ }	$3p_x$	$-1/9R$	12.5 a_0
3	1	−1	$3p_{-1}$ }	$3p_y$	$-1/9R$	12.5 a_0
3	1	0	$3p_0$	$3p_z$	$-1/9R$	12.5 a_0
3	2	2	$3d_2$ }	$3d_{x^2-y^2}$	$-1/9R$	10.5 a_0
3	2	−2	$3d_{-2}$ }	$3d_{xy}$	$-1/9R$	10.5 a_0
3	2	1	$3d_1$ }	$3d_{xz}$	$-1/9R$	10.5 a_0
3	2	−1	$3d_{-1}$ }	$3d_{yz}$	$-1/9R$	10.5 a_0
3	2	0	$3d_0$	$3d_{z^2}$	$-1/9R$	10.5 a_0

354]. We do not have any reason in the rest of this book to discuss the f-orbitals, and hence we do not describe them. Table 3.3 summarizes all orbitals of the hydrogen atom up to $n = 4$.

The Ground State of the Hydrogen Atom

With all of the orbitals for the simple hydrogen atom that we have described along with their resulting energies, we should remind the student once again of our second energy principle given in Chapter 1; that is:

> The most stable state of the hydrogen atom is the state with the lowest potential energy. This state is known as the *ground state.*

The energies given in Table 3.3 are total energies and include, therefore, both potential and kinetic energy contributions. It should be obvious that the potential energy is lower if the electron is in the 1s orbital than the 2s orbital since our potential energy for the hydrogen atom is given by:

$$V = -\frac{e^2}{4\pi\epsilon_0 r} \tag{3.48}$$

and r (or more correctly, $\bar{r}$, the average value of r) gets larger as n increases. Thus the potential energy becomes lowest when r is small. The kinetic energy of the electron also changes as n changes, but in fact it always changes in the opposite direction and by half as much.[5] Therefore, we

[5]This relationship is known as the *virial theorem* and is discussed more fully in Chapter 5.

Table 3.3 Summary of hydrogen-atom orbitals

n	l	m_l	Orbital	E [eV]	Number of orbitals = n^2
1	0	0	1s	−13.60	1
2	0	0	2s	−3.40	4
	1	+1, 0, −1	2p		
3	0	0	3s	−1.51	9
	1	+1, 0, −1	3p		
	2	+2, +1, 0, −1, −2	3d		
4	0	0	4s	−0.85	16
	1	+1, 0, −1	4p		
	2	+2, +1, 0, −1, −2	4d		
	3	+3, +2, +1, 0, −1, −2, −3	4f		

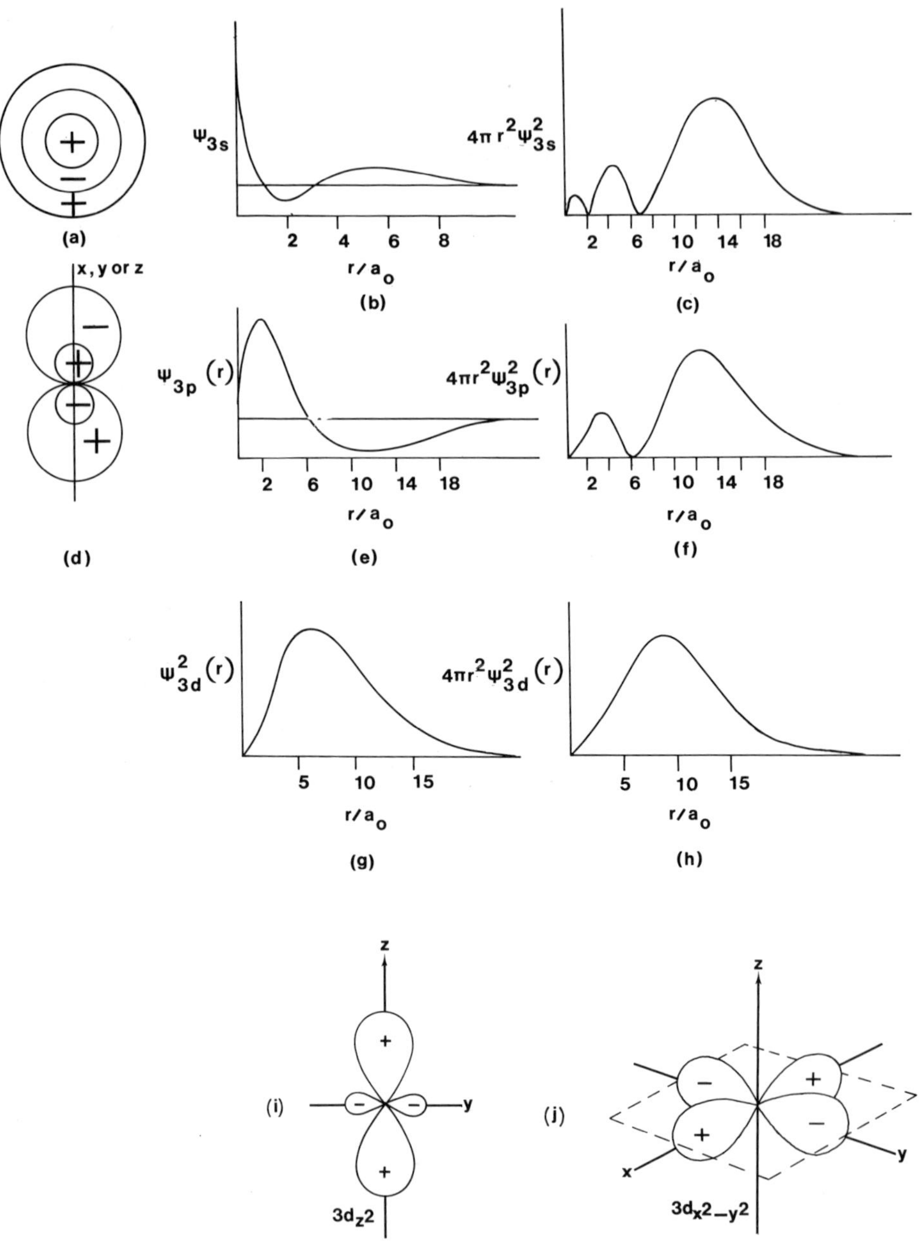

FIGURE 3.21. *(Above and facing page.)* The ψ_{3s} and a ψ_{3p} orbitals showing angular dependence (a and d), radial distribution (b and e), and radial dependence probability distribution (c and f). The ψ_{3d} orbitals—radial dependence (g), radial dependence probability distribution (h), angular dependence (i, j, k, l, m) and three-dimensional plots showing a contour of constant $|\psi|$ (n, o, p, q, r).

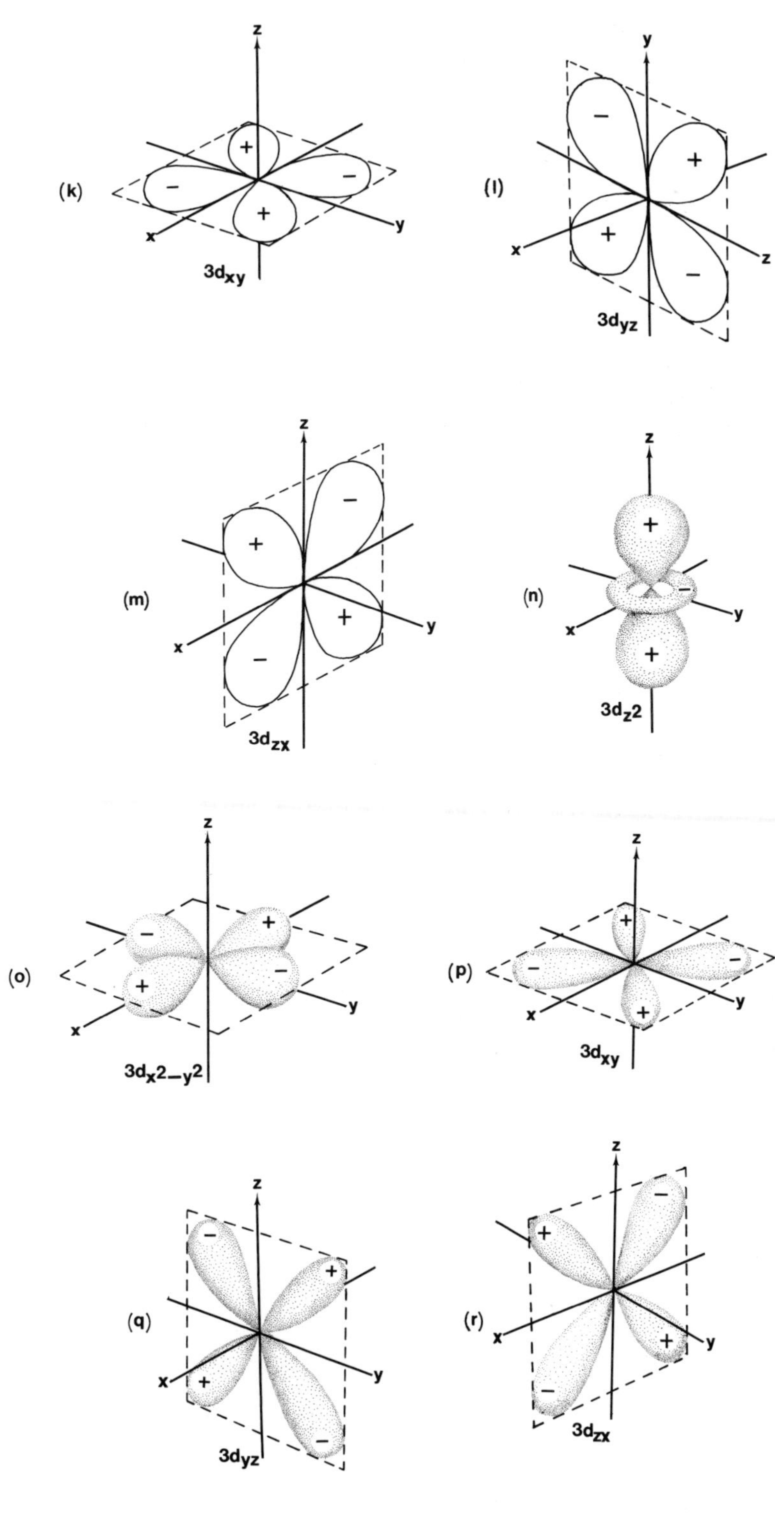
(k)
z
x
y
+
−
−
+
3d_xy
(l)
y
x
z
−
+
+
−
3d_yz
(m)
z
x
y
+
−
+
−
3d_zx
(n)
z
x
y
+
−
+
3d_z2
(o)
z
x
y
−
+
+
−
3d_x2−y2
(p)
z
x
y
+
−
−
+
3d_xy
(q)
z
x
y
−
+
+
−
3d_yz
(r)
z
x
y
+
−
−
+
3d_zx

can say that the lowest energy state is also the lowest potential-energy state for the hydrogen atom. So our single electron resides happily in the 1s orbital unless it can somehow gain energy externally through the interaction with light, such as occurs in an absorption spectrum. When the electron is in a higher energy orbital, the hydrogen atom is said to be in an *excited state*. Excited states of hydrogen are *not stable* because they are not the states of lowest potential energy. A hydrogen atom in an excited state will quickly lose the discrete excess energy between the excited state and lower states and in the process eventually drop back to the most stable energy configuration, namely, the ground state. This may give rise to an emission spectrum of the hydrogen atom.

Other One-electron Atoms

Singly ionized helium, He^+, and doubly ionized lithium, Li^{2+}, are also examples of one-electron atoms. The Schroedinger equation is identical to that for the hydrogen atom except that the additional nuclear charge (2 for He^+, +3 for Li^{2+}) has to be taken into account.

From the energy expression

$$E = -Z^2 \frac{R}{n^2}. \tag{3.49}$$

This means that each orbital will be four times lower in energy for He^+ and nine times lower in energy for Li^{2+}.

Similarly, for the expression giving us the average distance $\bar{r}$ of the electron from the nucleus

$$\bar{r} = \frac{a_0 n^2}{Z}\left[\frac{3}{2} - \frac{l(l+1)}{2n^2}\right] \tag{3.50}$$

the inverse Z dependence means that the orbitals get progressively smaller going from H to He^+ to Li^{2+}. This is an entirely expected result because the increased positive charge on the nucleus should attract the negative electron more strongly.

Summary

This has been a long chapter. We have traced the theories evolved by physicists and chemists to arrive at the present-day theory of the atom. In so doing we see that the simplest atom, the hydrogen atom, is described using the quantum mechanics developed by Planck, de Broglie, Schroedinger, Heisenberg, and others. The basis of this theory involves recognition of the wavelike characteristics of matter where restraints in the form of boundary conditions automatically give rise to the discrete nature of the quantum.

For the hydrogen atom, description of mechanical properties such as one-electron energies and average distance from nucleus are given in terms of three quantum numbers, n, l, and m_l. We depict the discrete quantum levels in which the electron moves in terms of orbitals that are mathematical functions describing the state of the electron and in terms of their squares, telling us where the electron is. We further have seen that the theory works for all one-electron systems, accurately predicting their properties as well.

It now behooves us to carry quantum mechanics to more complicated systems, namely: those atoms with more than one electron and molecules containing more than one nucleus. These topics are treated next in Chapters 4 and 5.

Problems

3.1 Calculate the energy equivalent in kJ mol^{-1} of:
a. Radiowaves of wavelength 10^4 m
b. X-rays of wavelength 154 pm

3.2 Calculate the hypothetical mass of a quantum in the X-ray region of wavelength 154 pm.

3.3 Express 1 cm^{-1} in its equivalent of frequency (Hz), wavelength (nm), energy (J mol^{-1}), and energy (eV).

3.4 What is the wavelength (in nm) associated with a photon of energy 10^{-17} J?

3.5 What is the wavelength associated with an electron moving with a velocity of 0.1 c?

3.6 Derive a formula analogous to Equation 3.14 for the Lyman lines in the spectrum of atomic hydrogen.

3.7 Calculate the radius of the second Bohr orbit.

3.8 Draw the third harmonic vibrations of a drum head. Show the nodes and antinodes.

3.9 List the names and allowed values of the quantum numbers n, l, and m_l.

3.10 How many f-orbitals are there for a given value of the n quantum number?

3.11 Show how the energy and the average distance from the nucleus of an electron in a hydrogen-like atom depend on the atomic number.

3.12 Sketch graphs for a 1s atomic orbital to show how the following vary with r:

a. ψ_{1s}
b. $|\psi_{1s}|^2$
c. $4\pi r^2|\psi_{1s}|^2$

3.13 Illustrate by means of a labeled diagram a $2p_z$, $3p_z$, and $3d_{x^2-y^2}$ atomic orbital.

3.14 Compare the energy of a single electron in a $3p_x$ atomic orbital for H and Li^{2+}.

4 Many-electron Atoms

We have seen in the last chapter that the energy of an electron in a one-electron atom or ion is given by Equation 3.49 as:

$$E_n = -Z^2 \frac{R}{n^2}.$$

For a given integer n, the energy decreases as the square of the atomic number (or charge on the nucleus) increases, and for given atomic number there is an array of energy levels available similar in number to that for hydrogen given in Figure 3.6, but with a different energy scale for each different value of the atomic number. The other important feature in the description of one-electron systems is that when Z increases, the size of the orbital decreases, the decrease being approximately inversely proportional to the nuclear charge.

For many-electron atoms direct solutions to the Schroedinger equation such as we wrote down for the hydrogen atom and other one-electron ions are not possible because of electron–electron interaction terms. Nonetheless, using approximations we can come very close to the answers that give the atom's properties. So close that we are absolutely sure the Schroedinger equation, or quantum mechanics, is the correct way to proceed. The calculations point out that the knowledge we gleaned from the hydrogen atom in reference to atomic orbitals is sufficient for an approximation that would vastly simplify our treatment of multielectron atoms:

The atomic orbitals of the hydrogen atom, 1s, 2s, $2p_x$, $2p_y$, $2p_z$, 3s, $3p_x$, $3p_y$, $3p_z$, $3d_{yz}$, $3d_{xz}$, $3d_{xy}$, $3d_{x^2-y^2}$, $3d_{z^2}$, and so on, modified in size and energy to allow for the increased nuclear charge and the presence of other electrons, can be used to describe many-electron atoms.

There are many experimental reasons why a description of atoms with more than one electron should follow the exact description of hydrogen and its excited states, by describing these atoms in a first approximation by means of hydrogen-like atomic orbitals. The most convincing of these is to compare the orbital energy-level diagram of hydrogen with the periodic table, derived in the first place completely from comparisons of chemical and physical properties of the elements.

The correlation is shown in Figure 4.1. It is evident that the periodic table has twice as many elements as the number of orbitals available to the electron in the hydrogen atom for each value of n. The 1:2 correlation, as we see shortly, is a direct result of an extremely important principle known as the *Pauli principle* which dictates that each orbital can contain only two electrons. So, many-electron atoms are described by assigning electrons to hydrogen-like atomic orbitals with no more than two electrons in each of these. This building up of many-electron atoms is known as the *Aufbau principle*.

The electronic structure of the elements of the periodic table is built up by adding electrons to hydrogen-like orbitals one by one and putting no more than two electrons in each orbital.

Another correlation with the periodic table is that elements above each other in the table (and with similar chemical properties) have the same outermost electron

no. of orbitals	H atom orbitals (up to n=3)
9	3s $3p_x$ $3p_y$ $3p_z$ $3d_{xy}$ $3d_{xz}$ $3d_{yz}$ $3d_{z^2}$ $3d_{x^2-y^2}$ →
4	2s $2p_x$ $2p_y$ $2p_z$ →
1	1s →

FIGURE 4.1. Correlation between H-atom orbitals and the Periodic Table.

configurations except for the value of the principal quantum number, n. For example:

Element	Z	Configuration
Lithium	3	$(1s)^2(2s)^1$
Sodium	11	$(1s)^2(2s)^2(2p)^6(3s)^1$.

The electron configuration designated $(1s)^2(2s)^1$ means that there are two electrons in the 1s orbital and one electron in the 2s orbital. It is the outermost electrons, $(2s)^1$ and $(3s)^1$ that determine the chemical properties. In this case, as the electrons are both s, these properties are similar.

It is often convenient to group the orbitals of many-electron atoms in *shells* and *subshells*. Subshells have the same value of the l quantum number for a given n quantum number. Shells have similar energy, but not necessarily the same value of n. Thus 1s, 2s, 2p, 3s, 3p, 3d, and so on are each called a *subshell*. The shells are commonly[1] designated

$(1s)^2$	K-shell
$(2s)^2(2p)^6$	L-shell
$(3s)^2(3p)^6$	M-shell
$(4s)^2(3d)^{10}(4p)^6$	N-shell
$(5s)^2(4d)^{10}(4p)^6$	O-shell
$(6s)^2(4f)^{14}(5d)^{10}(6p)^6$	P-shell

and give rise to the familiar, 2,8,8,18,18,32, etc. rows in the structure of the periodic table. The electrons in the outermost shell are called *valence electrons*.

[1]We say "commonly" because for inner shells of elements with high Z the principal quantum number, n, generally designates the shell (see Figure 4.10).

periodic table

no. of elements

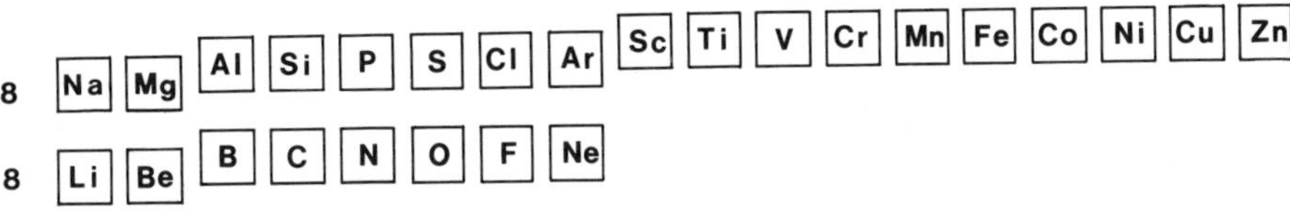

The rest of this chapter describes how the energies and sizes of the orbitals change from atom to atom and the effect of these changes on various physical and chemical properties.

Experiments Leading to Determination of Energies of Electrons in Atoms

As for hydrogen and one-electron ions, energy levels of many-electron atoms can be studied by their absorption and emission spectra. These studies reveal that the energy-level schemes are much more complicated than for one-electron species, but the lines in the spectra are again sharp and at least for some atoms the spectra show series of lines that may be interpreted in a similar, but not quite so simple, way as the series of lines in the hydrogen-atom spectrum (see Equations 3.17 and 3.18). Just as for hydrogen, the other spectra could be interpreted by assuming that the lines corresponded to the energy of the atom being changed in a way such that a single electron was being promoted from an orbital of one discrete energy to an orbital of another discrete energy. To illustrate this, Figure 4.2 shows part of the spectrum of the sodium atom and Figure 4.3 shows a partial interpretation of this spectrum in terms of an energy-level diagram. Obviously in the energy ranges that are easy to study, the spectra provide very much less than complete information about all possible energy levels of the atom. The significance of the labels in the energy-level diagram is seen later in this chapter. The spectra involving only the outermost electrons hopping from one energy level to another are called *optical spectra,* because the energy changes involved occur in the optical (visible, ultraviolet, and near infrared) region of the electromagnetic spectrum.

Other important information can be obtained by studying X-ray spectra. There are two ways of doing this. An atom can be bombarded with electrons with energies sufficient to remove inner or more tightly bound electrons

FIGURE 4.2. Part of the absorption spectrum of sodium vapor. These lines correspond to the nP ← 3S transitions shown in Figure 4.3.

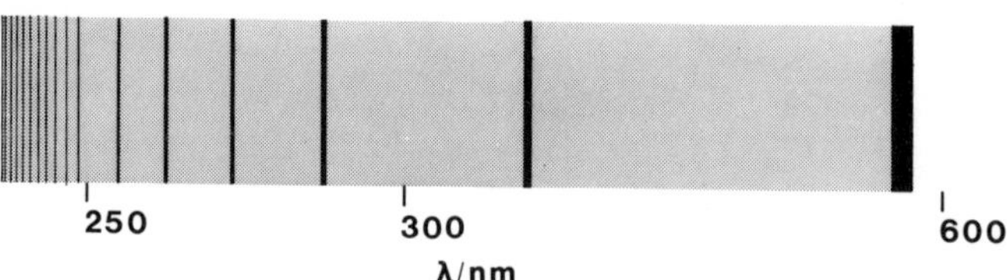

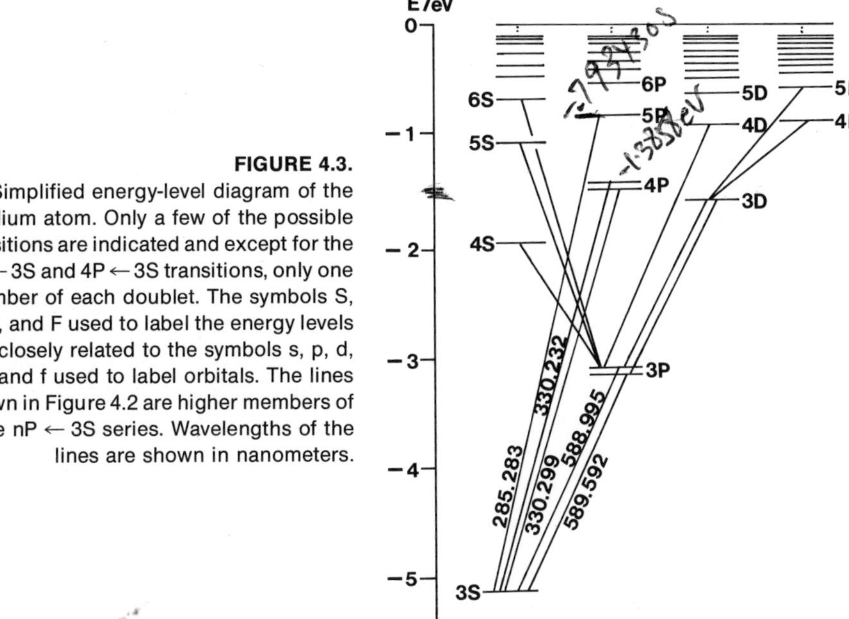

FIGURE 4.3. Simplified energy-level diagram of the sodium atom. Only a few of the possible transitions are indicated and except for the 3P ← 3S and 4P ← 3S transitions, only one member of each doublet. The symbols S, P, D, and F used to label the energy levels are closely related to the symbols s, p, d, and f used to label orbitals. The lines shown in Figure 4.2 are higher members of the nP ← 3S series. Wavelengths of the lines are shown in nanometers.

rather than the optical or outermost ones. At certain electron energies inelastic collisions corresponding to ionization of the inner electrons occur. The process is shown schematically in Figure 4.4, and we see that values of 1080.24 eV, 70.75 eV, and 38.09 eV are found for sodium.

FIGURE 4.4. X-Ray spectroscopy. Sodium atoms bombarded by high-energy electrons are ionized at distinct values of the energy of the bombarding electrons corresponding to the orbital energies.

1080.24 eV electron → Na → 1s electron + Na^+

$(1s)^1(2s)^2(2p)^6(3s)^1$

70.75 eV electron → Na → 2s electron + Na^+

$(1s)^2(2s)^1(2p)^6(3s)^1$

38.09 eV electron → Na → 2p electron + Na^+

$(1s)^2(2s)^2(2p)^5(3s)^1$

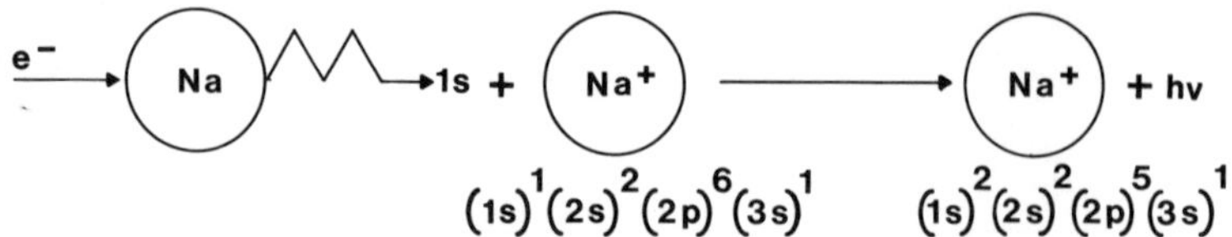

FIGURE 4.5. X-Ray spectroscopy. Sodium atoms bombarded by high-energy electrons. Measurement of orbital energies is made via photons ejected from sodium ions. In this case the emitted photon carries the orbital energy difference corresponding to the electron transition 2p → 1s.

These correspond to the energies required to ionize the 1s, 2s, and 2p electrons in sodium.[2] They correspond to orbital energies of these orbitals. The other way of observing X-ray spectra is shown in Figure 4.5 Here after bombarding sodium, for example, with electrons possessing energy sufficient to ionize a 1s electron, the electrons in higher energy orbitals fall down to fill its place and release the energy difference between the two levels as light photons.

The other useful technique for gaining information concerning energy levels of electrons in atoms is a direct measurement of ionization potentials. The energy required to remove the most loosely bound electron from an atom is termed the *first ionization potential* (I_1). That required to remove the next one from the monopositive ion is the *second ionization potential* (I_2), and so on. It is important to note the difference between the ionization potentials and the orbital energies derived from X-ray spectra. The former involved removing electrons from ions of increasing positive charge as we remove electrons of lower energy, while the latter involve removing the same electrons but always from the neutral atom. These then give the orbital energies, which are of smaller magnitude than the ionization potentials for the same orbital. This is illustrated later in Figures 4.8 and 4.9.

Electron Spin and the Pauli Principle

From a study of the effects of magnetic fields on the spectra of atoms there was evidence that some additional angular momentum should be associated with an electron and with it an additional quantum number. The fact that many atomic spectral lines, on close examination, form not a single line but two or more was also evidence of

[2]It may be well to point out that we really cannot say that we ionize a 1s electron (electrons are indistinguishable), but it is the 1s electron that is missing in the ion after ionization.

some additional quantum effect. (The best known example is the D-line of sodium at 589 nm, which corresponds to excitation of an electron from a 3s to a 3p orbital. The "line" is actually two lines of wavelengths 588.995 nm and 589.592 nm.) The fact that there is an additional quantization associated with electrons was particularly well indicated by experiments performed by O. Stern and W. Gerlach (1922). The experiments involved passing beams of neutral silver atoms with zero orbital angular momentum[3] through an inhomogeneous magnetic field as illustrated in Figure 4.6. The result of this experiment was that the beam was split into two equal beams, one bent toward the direction of the field and one away from it. It remained for S. Goudsmit and G. Uhlenbeck (1926) to interpret this result. Their hypothesis was that the electron had an intrinsic angular momentum in addition to orbital angular momentum described by the quantum number l. This means that in addition to properties described as due to a charged point particle, the electron must also have properties due to nonspherical distribution of charge and mass within its small size. They described the intrinsic angular momentum as *spin* and postulated two additional quantum numbers to describe it.

s—termed the *spin angular momentum quantum number*. It can have only one value $= \frac{1}{2}$ for a given electron regardless of the value of any other quantum number.

[3]A silver atom has a single electron in an s orbital. This electron has zero orbital angular momentum, that is, $l = 0$. All of the other electrons occupy orbitals in such a way that their angular momenta cancel.

FIGURE 4.6. The Stern–Gerlach experiment.

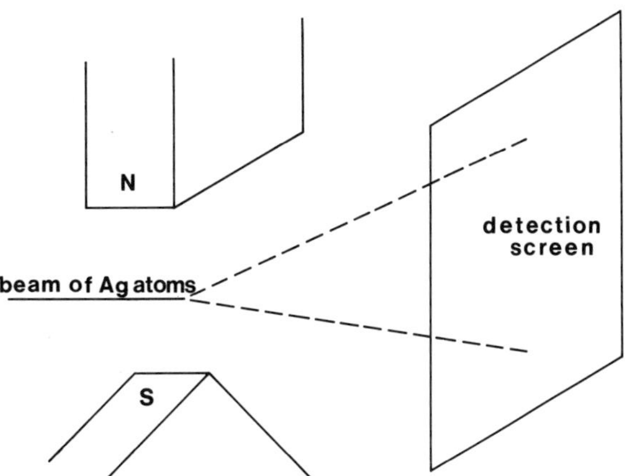

m_s—termed the *spin quantum number*. It represents a component of the spin angular momentum in a specified direction. It can have values from $+s$ to $-s$ differing by one. This makes the values of m_s trivial in that for a single electron they can only be $+\frac{1}{2}$ and $-\frac{1}{2}$.

These quantum numbers are analogous in meaning to l and m_l, the orbital angular momentum quantum numbers; s is the spin angular momentum quantum number, and m_s gives its value in a specified direction, l is the orbital angular momentum quantum number, and m_l gives its value in a specified direction.

In the Stern–Gerlach experiment the two beams correspond to the electrons in the original beam having half with $m_s = +\frac{1}{2}$ and half with $m_s = -\frac{1}{2}$. In one case the spin angular momentum is oriented toward the direction of the field, in the other against it. The spinning electron behaves like a small bar magnet, hence one beam of electrons is attracted towards the field of the magnet and the other repelled.

The hypothesis of electron spin angular momentum, and the quantum numbers s and m_s that go with it, has enabled scientists to account quite simply for many experimental observations on atoms and molecules. The spin property emerges directly from the relativistic quantum treatment of the hydrogen atom. Finally, and importantly for our purposes, it also leads to a simple way of stating the *Pauli principle*.

> No two electrons in an atom can have the same values for all four quantum numbers, n, l, m_l, and m_s.

The quantum number s is always equal to $\frac{1}{2}$; hence it is redundant to include it in our statement of the Pauli principle.

Now we recall that atomic orbitals are specified by the different possible sets of values for n, l, and m_l. For example, the quantum numbers of a $2p_z$ orbital are:

$$n = 2$$
$$l = 1$$
$$m_l = 0.$$

This means that only two electrons can be put into this or any other orbital, one with $m_s = \frac{1}{2}$ and one with $m_s = -\frac{1}{2}$. These two values can be represented in different ways:

$$m_s = \tfrac{1}{2} \text{ represented as } \alpha \text{ or } \uparrow$$
$$m_s = -\tfrac{1}{2} \text{ represented as } \beta \text{ or } \downarrow$$

The arrow symbolism is particularly convenient in describing the electronic structure of many - electron atoms. When two electrons are in the same orbital (we represent the orbital as a box, □), the spins can be represented as

and they are said to be *paired* or *antiparallel.* When two electrons are in different orbitals, the spins can be like this:

↑ ↑

and they are described as *unpaired* and *parallel.* Alternatively, they could be like this:

↑ ↓

and are described as *unpaired* and *antiparallel.* An important generalization known as *Hund's rule* has been discovered. It applies when there is a choice of putting electrons into orbitals of equal energy, that is, either in different orbitals with paired and antiparallel spins, or unpaired and parallel spins, or in the same orbital with paired spins.

> The configuration that gives the lowest energy state is that in which the electrons are in different degenerate orbitals with parallel spins.

Energy Levels of Many-electron Atoms

We are now ready to look at some of the details of the energies of electrons in many-electron atoms, having established that we will use the Aufbau principle, the Pauli principle, and Hund's rule. Before enunciating this in full it is most important to realize that we do this as a first approximation since even for atoms of low Z and especially for atoms of high Z, the Schroedinger equation cannot be solved exactly. For the hydrogen atom the potential energy of the electron consists of just one term, namely, its electrostatic attraction to the nucleus. For helium there are two electrons and hence attractions of both to the nucleus as well as the electron–electron repulsion contribute to the potential energy. This complexity of the potential energy is already such that it is exceedingly difficult to solve the Schroedinger equation in a precise manner, and when we get to atoms with many more electrons the situation becomes much worse. What we do is consider each electron separately as we feed them into orbitals. Then we estimate in a qualitative fashion the effect of the electron–electron repulsion and consider the fact that each electron will shield[4] the others from the positive nucleus to some extent. The details and general principles are best shown if we consider in detail the first 10 elements of the periodic table.

Helium, $Z = 2$

A useful way to consider the electronic structure of the helium atom is to use the readily available values of the energies required to completely remove, one after the other, the two electrons, that is, the first and second ionization potentials I_1 and I_2, respectively. The first ionization potential of He is 24.580 eV and the second ionization potential (i.e., the first of He^+) is 54.380 eV, four times that of the hydrogen atom (see Equation 3.49). If we build up the electronic structure of the He atom by putting the first electron into the 1s orbital, we have the He^+ ion with electron energy four times lower than that of the hydrogen atom. When we place the second electron into the 1s orbital, its energy will *not* be 54.4 eV for two reasons:

1. The second electron will be repelled by the first, raising the energy (electron–electron repulsion).
2. The second electron will be shielded somewhat by the first from the attractive force of the nucleus, so that the effective nuclear charge of the orbital is less than 2 (approximate calculations of this effect for helium imply that $Z' = 2 - 0.375 = 1.625$ for the 1s orbital, i.e., that the shielding effect of other electron is 0.375). As a result, $\bar{r}$ is increased from 0.75 a_0 for He^+ to 0.92 a_0 for He, and the corresponding orbital energy of He is raised from its value for He^+. Of course this is a mutual effect—the first electron equally is shielded from the nucleus by the second. Both occupy the expanded 1s orbital that corresponds to an effective nuclear charge of less than 2.

[4]For an electron that happens to be at some distance from the nucleus, the average repulsive effect of other electrons nearer to the nucleus is equivalent to having a lower positive charge on the nucleus. We say that the inner electrons "shield" the outer electrons from the nucleus. As a consequence the outer electrons occupy orbitals of a size (i.e., with a value of $\bar{r}$; see Equation 3.50) corresponding to a reduced value of Z, a value Z' called the *effective nuclear charge*.

$$Z' = Z - \sigma$$

where σ is *the shielding effect* of other electrons in the atom.

Both of these effects are rather difficult to calculate precisely because the motion of the electrons is correlated to some extent—there will be higher probability for them being on different sides of the nucleus than on the same side. It is well to remember at this stage that we cannot distinguish between electrons in an atom, so that the discussion above is a bit unrealistic except that it is the reverse of the ionization processes for which we can get the experimental numbers.

$$\begin{aligned} \text{He} &\rightarrow \text{He}^+ + e^- \quad & I_1 &= 24.6 \text{ eV} \\ \text{He}^+ &\rightarrow \text{He}^{++} + e^- \quad & I_2 &= 54.4 \text{ eV}. \end{aligned}$$

Let us now consider the excited electron configurations of He, $(1s)^1(2s)^1$ and $(1s)^1(2p)^1$. If we recall for a moment the energies of the hydrogen atom, the ionization potential when the electron is in the 2s orbital or the 2p orbital is the same because these orbitals are degenerate in energy:

$$\begin{aligned} \text{H}(2s)^1 \quad & I_1 = 3.4 \text{ eV} \\ \text{H}(2p)^1 \quad & I_1 = 3.4 \text{ eV}. \end{aligned}$$

Now for helium we find for ionizing an electron out of the 2s or 2p orbital that different energies are required:

$$\begin{aligned} \text{He}(1s)^1(2s)^1 \quad & I_1 = 4.36 \text{ eV} \\ \text{He}(1s)^1(2p)^1 \quad & I_1 = 3.49 \text{ eV}. \end{aligned}$$

This is due to the fact that there is a greater penetration of the core of the atom (the nucleus plus the 1s electron) by the 2s orbital than the 2p orbital. The electron distributions for the two helium configurations $(1s)^1(2s)^1$ and $(1s)^1(2p)^1$ are such that the effective nuclear charge Z' (2s) is greater than the effective nuclear charge Z' (2p). Thus we expect the 2s orbital to have a lower energy than the 2p orbital in helium, which is in agreement with the experimental data given above. This energy difference between orbitals with the same n but different l is extremely important in determining the structure of the periodic table. Figure 4.7 summarizes these energy relationships for He and compares them with those of H.

Lithium, $Z = 3$

For lithium, because of the Pauli principle, two electrons will go into the 1s orbital and the third will go into the lowest energy vacant orbital available. As for helium, the 2s orbital for lithium is lower in energy than the 2p, so that the lowest energy configuration of lithium is $(1s)^2(2s)^1$. The quantum numbers for these three electrons are:

Electron	1	2	3
n	1	1	2
l	0	0	0
m_l	0	0	0
m_s	½	−½	½ (or −½).

The ionization potentials of the $(1s)^2(2s)^1$ and the $(1s)^2(2p)^1$ configurations of lithium are 5.390 eV and 3.542 eV, respectively, further justifying our placing the 2s orbital lower in energy than the 2p.

It is also interesting to look at some of the other excited electron-configuration ionization potentials. For the $(1s)^2(3s)^1$, $(1s)^2(3p)^1$, and $(1s)^2(3d)^1$ configurations, they are 2.017 eV, 1.556 eV, and 1.512 eV, respectively. Again we see that the levels with the same n but different l values have different energies increasing in the order s < p < d. This ordering is quite general for many-electron atoms.

Beryllium to Neon

We have now established the general principles of building up the electron configurations of atoms. We feed electrons into hydrogen-like atomic orbitals, starting with the lowest-energy ones and put two electrons in each orbital. For many-electron atoms, orbitals of a given n have energies in the order ns < np < nd, and so on. Hence the electron configurations of the elements from Li to Ne can be written down at once and are:

FIGURE 4.7. Orbital energy relationships for H and He.

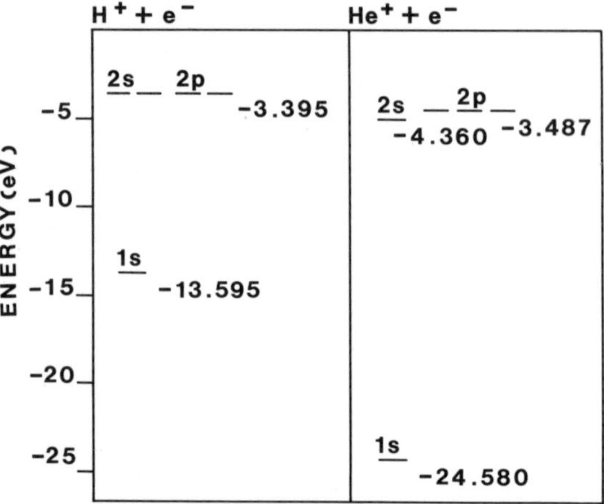

Li	$(1s)^2(2s)^1$
Be	$(1s)^2(2s)^2$
B	$(1s)^2(2s)^2(2p)^1$
C	$(1s)^2(2s)^2(2p)^2$
N	$(1s)^2(2s)^2(2p)^3$
O	$(1s)^2(2s)^2(2p)^4$
F	$(1s)^2(2s)^2(2p)^5$
Ne	$(1s)^2(2s)^2(2p)^6$.

Figure 4.8 shows the ionization potentials of these atoms on a logarithmic scale and Figure 4.9, the orbital energies of the different orbitals in these atoms based on X-ray spectral data. *Again we must remember that the succeeding ionization potentials of an atom are those for ions of increasing positive charge, whereas the orbital energies represent the energy required to remove the electron from the particular orbital in the neutral atom.* A number of general features of the energy variations can be deduced from a perusal of the figures.

1. There is a general decrease in orbital energy as Z increases for the 1s, 2s, and 2p orbitals. This is true also when Z increases further, but in addition, for higher Z some crossovers between orbitals of different principal quantum occur that are very important for the structure of the periodic table.
2. There is a large energy gap in both the ionization potentials and orbital energies when the principal quantum number of the electron involved is changed from $n = 1$ to $n = 2$. As we saw for hydrogen, there is a correlation between the energy of an orbital and the average distance, $\bar{r}$, of the orbital from the nucleus. This suggests the concept of shells of electrons in atoms with similar energies. These shells of electrons will have most of their electron density at a particular distance from the nucleus. Thus as we go from one shell to another such as from the K-shell (1s) to the L-shell (2s, 2p), the electrons in the outer shell are more loosely bound. In other words, they have higher energies and are more easily ionizable. Qualitatively we see that this is reasonable because the inner shell shields the nucleus from the outer shell, the electron–electron repulsion compensating for some of the nucleus–electron attraction for the electrons in the outer shell. Notice also that for the atoms Li through Ne the 1s orbital energy decreases substantially so that the 1s shell gets closer

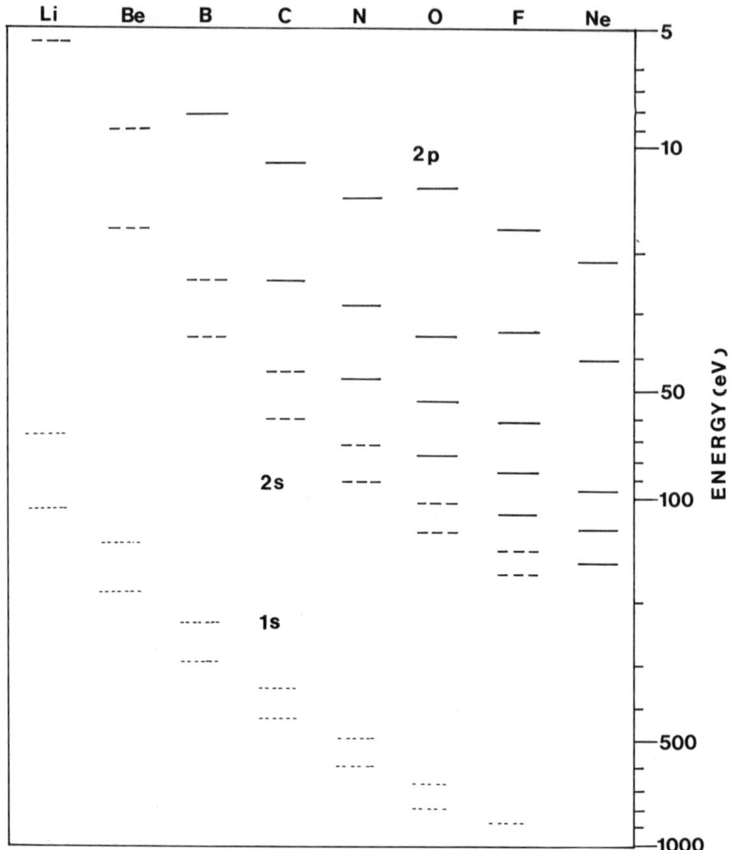

FIGURE 4.8. Ionization potentials of the atoms Li through Ne. Note that ionization potentials are positive quantities.

and closer to the nucleus as we go through this series. This explains why the size discrepancy between atoms of low Z and high Z is not as large as might have been expected.

3. Although there is a general increase in first ionization potentials when going from lithium to neon, there are two irregularities in this trend.
   a. The first ionization potential of boron is less than that of beryllium. This we can expect since the 2p orbital energy for B is higher than the 2s orbital energy in Be.
   b. The first ionization potential of O is less than that of N. Here the reason is due to Hund's rule in that when the 2p orbitals are being filled the lowest energy can be achieved if the electrons go into separate p-orbitals. This reduces the electron–electron repulsion since two electrons are less likely to

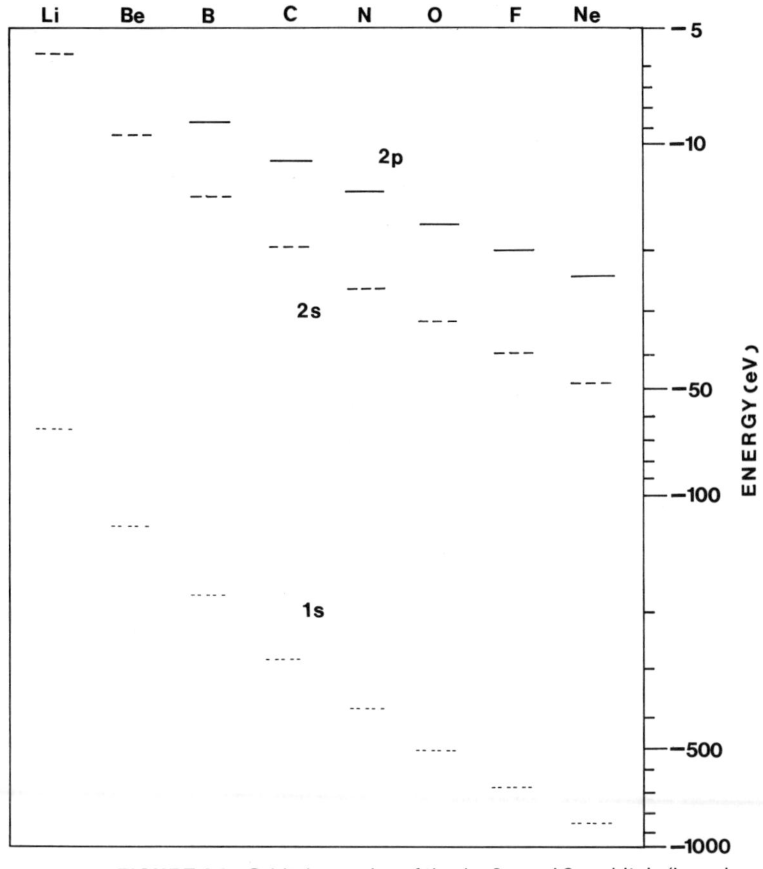

FIGURE 4.9. Orbital energies of the 1s, 2s, and 2p orbitals (based on X-ray spectral data) of the atoms Li through Ne. Note that the orbital energies are negative.

be close to each other if they occupy different orbitals than if they occupy the same orbital.

N : [↑] [↑] [↑]
$2p_x$ $2p_y$ $2p_z$

extra electron – electron repulsion → [↑↓] [↑] [↑]
O : $2p_x$ $2p_y$ $2p_z$

Thus the first electron to be ionized in oxygen is one of the paired electrons[5] and is easier to remove than

[5]See, however, footnote 2 of this chapter.

a 2p electron in nitrogen. These two cases are quite general and can be summarized. *For the outermost electrons, half-filled and completely filled subshells have special stability for the atom.*

From Sodium on

We have been rather detailed in the description of the first 10 elements in the periodic table. This has been to try to make clear exactly what the ionization potentials and orbital energies mean. Figure 4.10 shows how the orbital energies vary with Z for $Z = 1–100$. Because of the huge variation in energy, the ordinate in this figure is $(-E)^{1/2}$ and logarithmic. The 1s, 2s, and 2p orbital energies continue to show the same sort of variation with Z that we have described for the first 10 elements. After this, for $n > 3$, the orbitals for a given n still have higher energies for higher l values but, for certain values of Z, there are crossovers between orbitals with different values of n. The first of these occurs when Z is about 10, when the 4s orbital energy becomes less than the 3d, so that it fills up before the 3d. It should be noted from Figure 4.7 that for even higher values of Z (>30), the 3d orbital energy comes back below 4s, so that when Z has high values the inner shells of electrons are made up of those electrons all with the same principal quantum number. For orbitals above 3d in energy the crossings get quite complicated. The important ones for the Aufbau principle are listed below. The energy relationships given *are generally correct only when the number of electrons is equal to Z and the outermost shell is being filled.* The order is usually different for Z values either higher or lower than this. The crossovers are:

$$\begin{aligned} &3d > 4s \\ &4d > 5s \\ &5d > 4f > 6s \\ &6d > 5f > 7s. \end{aligned}$$

Figure 4.11 is an energy-level diagram, correct for the order of energy levels when the last electron in an atom goes into the orbital that one is looking at. This may be called the ‘‘filling-order’’ orbital energy-level diagram. Notice the shells of similar energies and that the gaps between shells occur after the p-orbitals are filled. The electron configurations corresponding to filling of the p-orbitals are those of the rare gases, and their relative

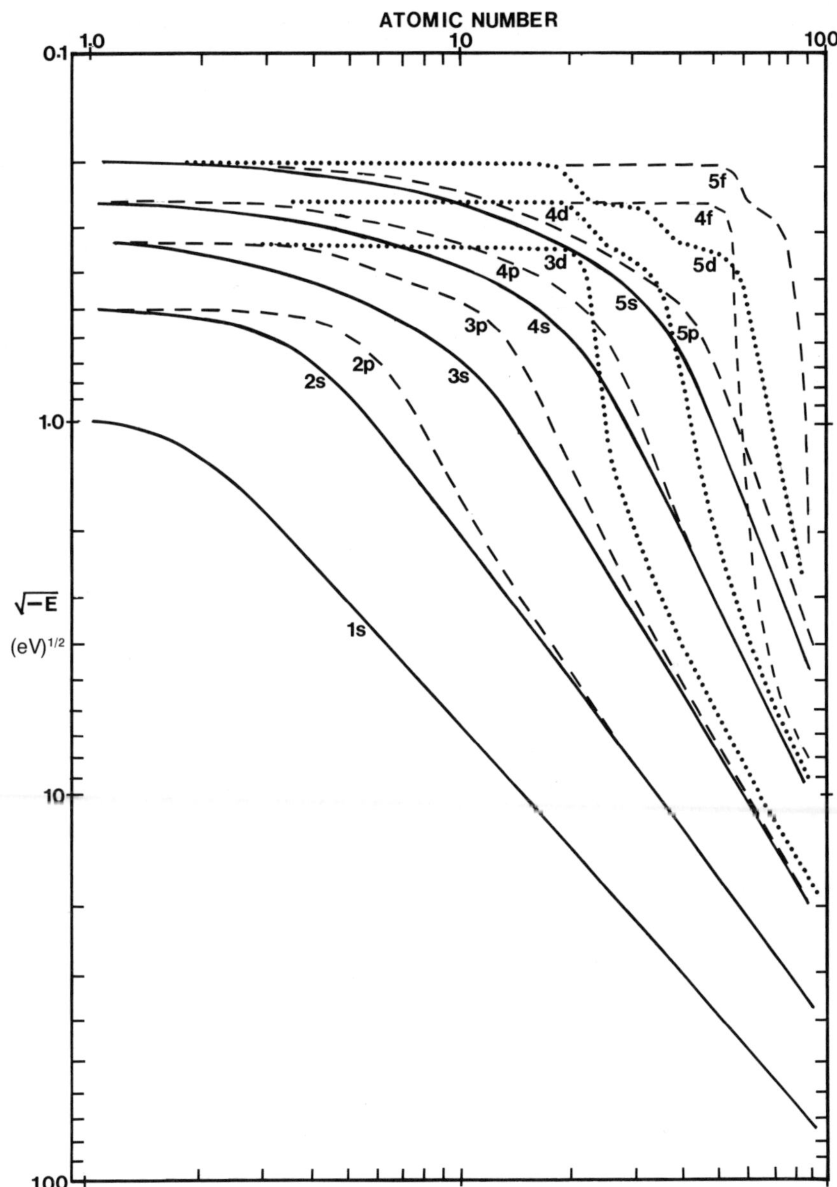

FIGURE 4.10. Orbital energies as a function of Z. The curves were obtained by approximate solution of the wave equation.

chemical inertness is related to the large energy gap to the next vacant orbital. An easy way to remember the orbital filling order is to write the orbitals for each principal quantum number in rows and draw arrows sloping from the right to the left diagonally as follows

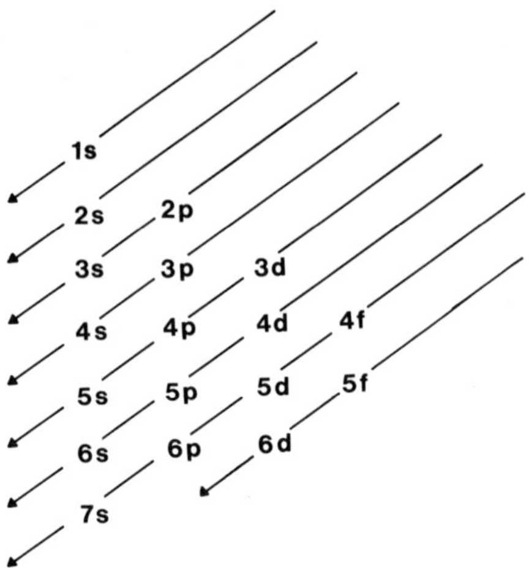

Taking the arrows in order from the top then gives the orbital filling order.

There are a number of irregularities in the lowest energy electron configurations of the elements. For copper we would assign the following electron configuration:

$$1s^2\ 2s^2\ 2p^6\ 3s^2\ 3p^6\ 3d^9\ 4s^2$$

but the configuration:

$$1s^2\ 2s^2\ 2p^6\ 3s^2\ 3p^6\ 3d^{10}\ 4s^1$$

gives a state with a slightly lower energy. Similarly, for chromium the most stable state comes from the configuration

$$1s^2\ 2s^2\ 2p^6\ 3s^2\ 3p^6\ 3d^5\ 4s^1$$

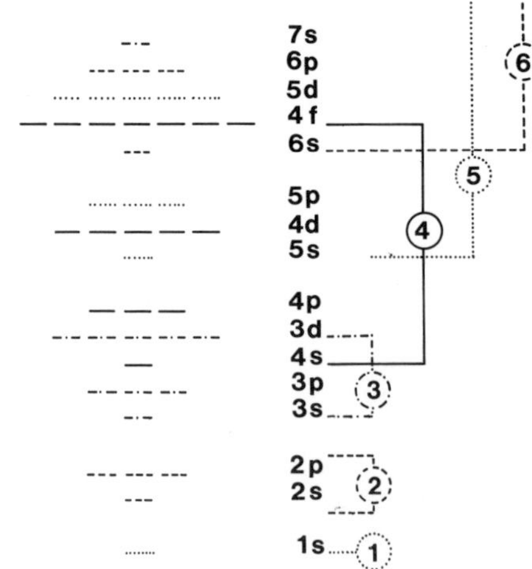

FIGURE 4.11.
Filling-order orbital energy-level diagram.

although the diagram would give

$$1s^2\ 2s^2\ 2p^6\ 3s^2\ 3p^6\ 3d^4\ 4s^2.$$

For higher values of Z there are more examples of this and mostly they are a result of our earlier generalization that filled and half-filled subshells of electrons lead to particularly stable states. For the half-filled cases it is a reflection of Hund's rule in that generally there is a maximum number of electrons with parallel spins in different orbitals.

The Periodic Table

When a complete list of electronic structures of the atoms is compiled it is evident that a recurring pattern occurs in the outermost, that is, most loosely bound electrons. One prominent feature is the filling of a set of s- and p-orbitals—often termed the *completion of an octet of electrons*. As we mentioned previously, the periodic variation in the nature of the outermost shell of electrons precisely reflects the variations of chemical and physical properties of the elements. These *valence electrons* play a central part in the theory of valency, as we see in Chapter 5. The concept of valence electrons is shown by listing the valence electrons of a few atoms in Table 4.1.

The systematic variation in the nature of the valence electrons in atoms of different atomic numbers is conveniently summarized in tabular form. Such a scheme is known as the periodic table of the elements. A particularly convenient form of the table is given inside the back cover.

A quick glance shows that one section of the periodic table consists of the main group of elements (the s-block and p-block) consisting of atoms with various numbers of s- and p-electrons as valence electrons, any other groups of degenerate orbitals (e.g., d- or f-orbitals) being either completely filled or empty. The various groups are numbered according to the number of valence electrons; for

Table 4.1 Valence electrons of some atoms

Atom	Valence electrons	Number
H	$(1s)$	1
Li	$(2s)$	1
C	$(2s)^2(2p)^2$	4
F	$(2s)^2(2p)^5$	7
Ne	$(2s)^2(2p)^6$	8
Na	$(3s)$	1
Mg	$(3s)^2$	2
Ti	$(3d)^2(4s)^2$	4
Cr	$(3d)^5(4s)$	6

instance, the group I elements, the alkali metals, have a single valence electron (in an s orbital), the group IV elements have four valence electrons (in an s^2p^2 configuration), the halogens in group VII have seven valence electrons (s^2p^5), and so forth.

The remaining sections of the periodic table consist of the d-block, commonly referred to as the *transition elements,* and the f-block, comprising the lanthanides and actinides. The transition elements with a given number of d-electrons in their valence shell show some limited resemblance in their chemical properties to the main group elements having the same number of p electrons in their valence shell. For example, manganese, technetium, and rhenium (s^2d^5 elements) have some series of compounds analogous to the halogens (s^2p^5 elements). In the case of lanthanides and actinides, however, especially the former, their characteristic is a chemical similarity to one another rather than to main group elements.

Trends in Ionization Potential

Figure 4.12 shows how the first ionization potential of atoms varies with atomic number for the entire periodic table, and Table 4.2 lists the ionization potentials for the lower Z elements. As we have seen before, for the elements Li to Ne, there is a general increase in first ionization potential as we fill up the 2s and 2p orbitals. This

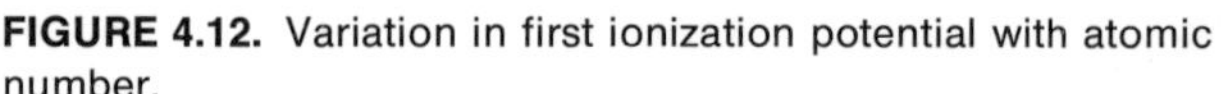
FIGURE 4.12. Variation in first ionization potential with atomic number.

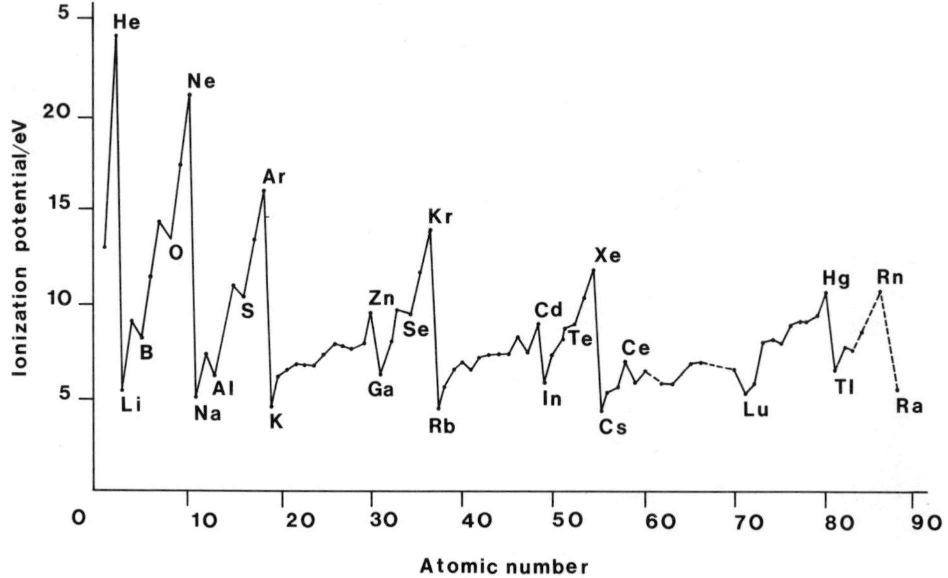

Table 4.2 Ionization potentials [in eV]

Element	Outer electron configuration	I_1	I_2	I_3	I_4	I_5	I_6	I_7	I_8
H	$1s^1$	13.60							
He	$1s^2$	24.58	54.40						
Li	$2s^1$	5.39	75.62	122.42					
Be	$2s^2$	9.32	18.21	153.85	217.66				
B	$2s^22p^1$	8.30	25.15	37.92	259.30	340.13			
C	$2s^22p^2$	11.26	24.38	47.86	64.48	391.99	489.84		
N	$2s^22p^3$	14.54	29.61	47.43	77.45	97.86	551.93	666.83	
O	$2s^22p^4$	13.61	35.15	54.93	77.39	113.87	138.08	739.11	871.12
F	$2s^22p^5$	17.42	34.98	62.65	87.23	114.21	157.117	185.14	953.60
Ne	$2s^22p^6$	21.56	41.07	64	97.16	126.4	157.91		
Na	$3s^1$	5.14	47.29	71.65	98.88	138.60	172.36	208.44	264.16
Mg	$3s^2$	7.64	15.03	80.12	109.29	141.23	186.86	225.31	265.96
Al	$3s^23p^1$	5.98	18.82	28.44	119.96	153.77	190.42	241.93	285.13
Si	$3s^23p^2$	8.15	16.34	33.46	45.13	166.73	205.11	246.41	303.87
P	$3s^23p^3$	11.0	19.65	30.16	51.35	65.01	220.41	263.31	309.26
S	$3s^23p^4$	10.36	23.4	35.0	47.29	72.5	88.03	280.99	328.80
Cl	$3s^23p^5$	13.01	23.80	39.90	53.5	67.8	96.7	114.27	348.3
Ar	$3s^23p^6$	15.76	27.62	40.90	59.79	75.0	91.3	124.0	143.46
K	$4s^1$	4.34	31.81	46	60.90		99.7	118	155
Ca	$4s^2$	6.11	11.87	51.21	67	84.39		128	147
Sc	$3d^14s^2$	6.56	12.89	24.75	73.9	92	111.1		159
Ti	$3d^24s^2$	6.83	13.63	28.14	43.24	99.8	120	140.8	
V	$3d^34s^2$	6.74	14.2	29.7	48	65.2	128.9	151	173.7
Cr	$3d^54s^1$	6.76	16.49	30.95	49.6	73	90.6	161.1	185
Mn	$3d^54s^2$	7.43	15.64	33.69		76	110.24	196	222
Fe	$3d^64s^2$	7.90	16.18	30.64					
Co	$3d^74s^2$	7.86	17.05	33.49					
Ni	$3d^84s^2$	7.63	18.15	36.16					
Cu	$3d^{10}4s^1$	7.72	20.29	36.83					
Zn	$3d^{10}4s^2$	9.39	17.96	39.70					
Ga	$4s^24p^1$	6.00	20.51	30.70	64.2				
Ge	$4s^24p^2$	7.88	15.93	34.21	45.7	93.4			
As	$4s^24p^3$	9.81	20.2	28.3	50.1	62.6	127.5		
Se	$4s^24p^4$	9.75	21.5	32.0	42.9	73.1	81.7	155	
Br	$4s^24p^5$	11.84	21.6	35.9					193
Kr	$4s^24p^6$	13.996	24.56	36.9					

occurs because these electrons are at about the same distance from the nucleus and thus do not shield each other very effectively from the nucleus. Hence the outermost electron on each succeeding element "sees" more positive charge from the nucleus and consequently is more tightly bound and difficult to ionize.

The same general increase occurs for the next row of elements as the 3s and 3p orbitals are filled up. Here the values for elements with similar outer electron configurations to the preceding shell are slightly less because the electrons with $n = 3$ are further from the nucleus than those with $n = 2$ and are almost completely shielded from the nucleus by the inner shells. The two irregularities again appear in that the ionization potential of the $(3s)^2$ configuration of Mg is greater than the $(3s)^2(3p)^1$ configuration of Al and that the $(3p)^3$ configuration of P is greater than the $(3p)^4$ configuration of S. These irregularities have been explained earlier (p. 73). The trends for the rest of the elements are similar in that for each row of the periodic table there is a general increase as we go across and for the same outer configurations there is a slight decrease due to

FIGURE 4.13. Plot of I_n/n against n for Be, B, C, and N.

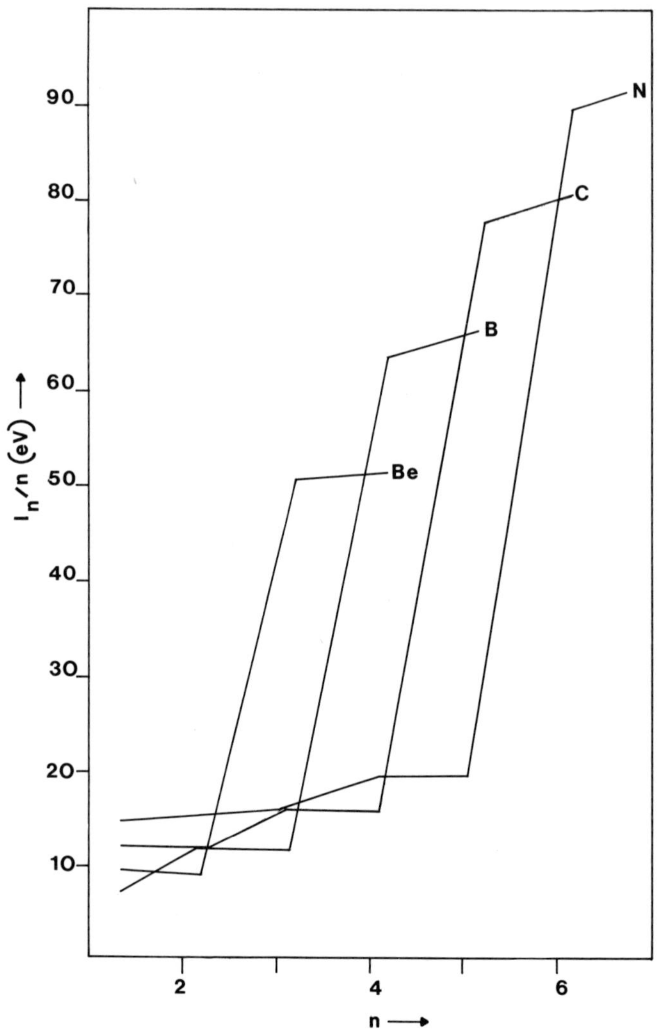

increasing distance from the nucleus. Again there are irregularities that occur after filling and half-filling subshells.

The big jump in successive ionization potentials when we get to the rare-gas electron configurations is nicely illustrated if we plot values of I_n/n against n where I_n is the nth ionization potential of an atom, so that n will be the charge on the ion produced by ionization in units of proton charge. Figures 4.13 and 4.14 show this for a number of atoms. In each case the large jump in I_n/n occurs when taking an electron from the $(ns)^2\ (np)^6$ configuration.

FIGURE 4.14. Plot of I_n/n against n for Na through Ar.

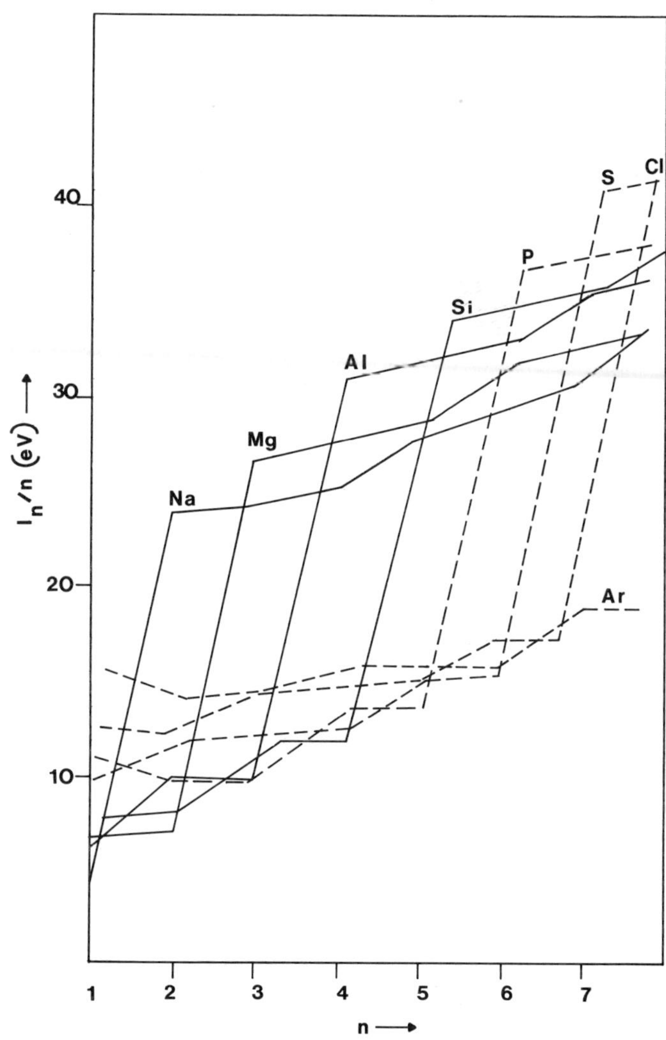

Table 4.3 Lowest energy configurations

Atom	Configuration	Ion	Configuration
Sc	$KLM(4s)^2(3d)^1$	Sc^{2+}	$KLM(3d)^1$
Ti	$KLM(4s)^2(3d)^2$	Ti^{2+}	$KLM(3d)^2$
V	$KLM(4s)^2(3d)^3$	V^{2+}	$KLM(3d)^3$
Cr	$KLM(4s)^1(3d)^5$	Cr^{2+}	$KLM(3d)^4$
Mn	$KLM(4s)^2(3d)^5$	Mn^{2+}	$KLM(3d)^5$
Fe	$KLM(4s)^2(3d)^6$	Fe^{2+}	$KLM(3d)^6$
Co	$KLM(4s)^2(3d)^7$	Co^{2+}	$KLM(3d)^7$
Ni	$KLM(4s)^2(3d)^8$	Ni^{2+}	$KLM(3d)^8$
Cu	$KLM(4s)^1(3d)^{10}$	Cu^{2+}	$KLM(3d)^9$
Zn	$KLM(4s)^2(3d)^{10}$	Zn^{2+}	$KLM(3d)^{10}$

Electron Configurations of Doubly Charged First-transition-series Ions

As we have seen, orbital energies vary considerably with the charge on the nucleus. Furthermore, the relative order of orbital energies changes in some cases, again depending on the nuclear charge. One consequence of this is that an *isoelectronic* (equal number of electrons) atom and ion may have different electron configurations in their lowest energy state.

To illustrate this, consider V and Mn^{2+}. Both have 23 electrons, but their electron configurations are:

$$\begin{array}{ll} \mathrm{V} & (1s)^2(2s)^2(2p)^6(3s)^2(3p)^6(3d)^3(4s)^2 \\ \mathrm{Mn}^{2+} & (1s)^2(2s)^2(2p)^6(3s)^2(3p)^6(3d)^5. \end{array}$$

The nuclear charge of V is +23, whereas that of Mn^{2+} is +25. The increased nuclear charge is just enough to make the $(3d)^5$ configuration of lower energy than $(3d)^3(4s)^2$.

This same reversal of orbital energies, that is, 4s and 3d, occurs for all of the doubly charged transition series ions so that there are no 4s electrons in their lowest energy configurations. Table 4.3 illustrates this, giving the lowest energy configuration for all of the first-transition metals and their doubly charged ions.

Finally, we note there is no general rule of thumb applicable to the monopositive ions. There is sometimes a 4s electron in the lowest energy configuration, and sometimes not.

Atomic Size

Atomic size in the orbital picture of electronic structure is necessarily a vague concept. But it is useful to know something about size variations between atoms. Experimentally we can get some idea at least of comparative atomic radii by taking half the bond distance in the diatomic molecule. For example, the bond distance in H_2 is 74 pm and so the atomic radius of hydrogen atom is 37 pm, and that for Li_2 is 267 pm, so that the atomic radius of lithium is 134 pm. Figure 4.15 gives a diagrammatic picture of the relative sizes of some atoms and atomic ions derived in this manner. Again there are trends that we can explain qualitatively.

To start with there is a regular decrease in size as we scan the horizontal rows in the periodic table. This arises because as the charge on the nucleus increases, and since the electrons added in the same shell do not shield each other from the nucleus very much, they are bound more firmly and tend to draw in closer to the nucleus. Then, to begin the next row, a loosely bound electron is added, which increases the size. But it is not that much greater than the element above it because of the decrease in size in going across the previous row.

For ions similar arguments hold. For positive ions, when the first s electron is removed the resulting positive ion is very much smaller than the netural atom. This inversely parallels the addition of this electron, which greatly increases the size of the atom from the element.

FIGURE 4.15. Atomic radii of some elements and their positive or negative ions. Radii are in picometers.

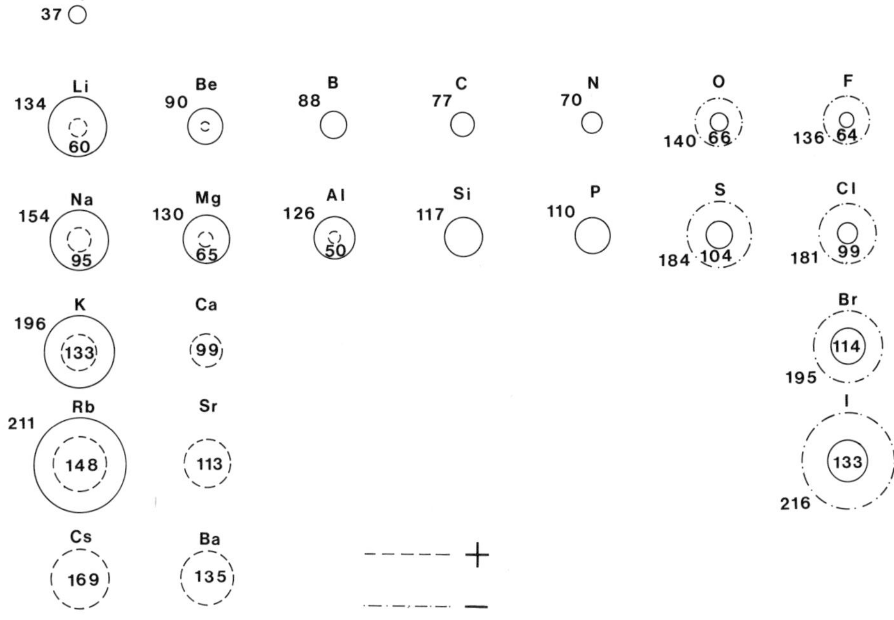

Similarly, the negative ions formed on adding an electron or electrons to fill the p-shell are greatly increased in size.

Electron Affinities

Another physical property that gives some insight into electronic properties of atoms is the electron affinity. The electron affinity (A) of an atom is the energy liberated when an electron is added to a neutral atom. Unfortunately, there are extreme experimental difficulties involved in obtaining accurate values of electron affinities. Some values are given in Table 4.4, mostly derived by indirect means. As with ionization potentials, it would be of value to have successive electron affinities, arising from the addition of further electrons to the negative ion. The values of these will always be negative and are only available for sulfur and oxygen. They are −8.9 eV and −6.1 eV for the second electron affinity of oxygen and sulfur, respectively.

Electron affinities together with ionization potentials give us a useful way of predicting the attraction one atom will have for the electron of another atom when forming molecules.

Table 4.4 Electron affinities [in eV]

Atom	Outer electron configuration	Electron affinities
H	$(1s)^1$	0.754
He	$(1s)^2$	−0.22
Li	$(2s)^1$	0.62
Be	$(2s)^2$	−2.5
B	$(2s)^2(2p)^1$	0.86
C	$(2s)^2(2p)^2$	1.27
N	$(2s)^2(2p)^3$	0.0
O	$(2s)^2(2p)^4$	1.465
F	$(2s)^2(2p)^5$	3.34
Ne	$(2s)^2(2p)^6$	−0.30
Na	$(3s)^1$	0.55
Mg	$(3s)^2$	−2.4
Al	$(3s)^2(3p)^1$	0.52
Si	$(3s)^2(3p)^2$	1.24
P	$(3s)^2(3p)^3$	0.77
S	$(3s)^2(3p)^4$	2.08
Cl	$(3s)^2(3p)^5$	3.61
Ar	$(3s)^2(3p)^6$	−0.36
K	$(4s)^1$	0.50
Br	$(4s)^2(4p)^5$	3.36
I	$(5s)^2(5p)^5$	3.06

Summary

In this chapter we have seen how we can qualitatively apply the quantum-mechanical description of the hydrogen atom to many-electron atoms. We are able to write down the electron configuration of any element in the periodic table, by introducing the concepts of: (a) electron spin, (b) the Pauli principle, (c) the Aufbau principle, and (d) Hund's rule.

Furthermore, experimentally determined properties of our many-electron atoms such as spectra, chemical periodicity, and trends in orbital energies, in ionization potentials, in atomic radii, and in electron affinities are readily explicable by using the orbital model.

We should feel confident at this stage to take our quantum mechanical atoms and allow them to form quantum-mechanical molecules. Again we must do this in a qualitative–pictorial fashion because actual solutions of the Schroedinger equation for molecules are mathematically intractable. Still, our foundation is firm and we present the subject of valency in Chapter 5 with confidence.

Problems

4.1 Describe:
a. the Pauli principle
b. the Aufbau principle
c. Hund's rule

4.2 What is the maximum number of orbitals for:
a. $n = 5$
b. $n = 4, l = 2$

4.3 Write down the "filling order" of atomic orbitals up to 6d.

4.4 Give the electronic configuration of:

a. P	e. Cr
b. Mn	f. Cr^{2+}
c. Te^{2-}	g. Cu
d. Gd	h. Co^{2+}

4.5 For each of the following sets, write down the atoms in order of *increasing* first ionization potentials:

a.	Be	N	F
b.	Be	Ca	Ba
c.	Be	B	C
d.	N	O	F
e.	F	Cl	Br
f.	Si	P	S

4.6 What is the eighth ionization potential of sodium? What is the orbital energy of a 2s electron of sodium? Why are these two quantities different?

4.7 What are the allowed values of the quantum numbers s and m_s?

4.8 What are the valence electrons for K, Fe, and Br?

4.9 How many unpaired electrons would you expect for the following in their lowest-energy electron configurations:
a. Co
b. Co^{2+}
c. Co^{3+}

4.10 Explain why cations and anions are smaller and larger, respectively, than their neutral atoms.

5 Molecular Theory and Chemical Bonds

The fundamental goal of the theory of valency is to explain chemical combination and the properties of chemical compounds in terms of the basic principles of physics and the properties of atoms and electrons.

Energy Changes and the Chemical Bond

Let us start our discussion by considering the changes in energy that might occur when two atoms, A and B, are moved toward each other. From our studies of electrostatics we know that if both atoms had lost an electron and become the positive ions A^+ and B^+, the potential energy would vary with R_{AB}, the distance between the atoms, in a pattern similar to that shown in Figure 5.1(a). We say that the two atomic ions are more stable when far apart than when in close proximity to each other because the lower the potential energy the more stable is the system.

If we are dealing with oppositely charged atomic ions A^+ and B^-, we have the reverse situation, as shown in Figure 5.1(b). The system of two ions is more stable when the ions are close together than when they are far apart.[1]

Finally, if we are dealing with two unchanged atoms, such that simple electrostatics gives us no obvious indication of how the potential energy will vary with atomic separation, more sophisticated study reveals the potential-energy story to be similar to that displayed in Figure 5.1(c). Down to an internuclear distance[2] of R_e the system

[1]In fact, because atomic ions have a finite size and cannot interpenetrate one another, the potential energy curve will ultimately rise rapidly at very small values of R_{AB}.

[2]The distance between two atoms is usually taken to be the distance between the two nuclei and so is termed the *internuclear distance*.

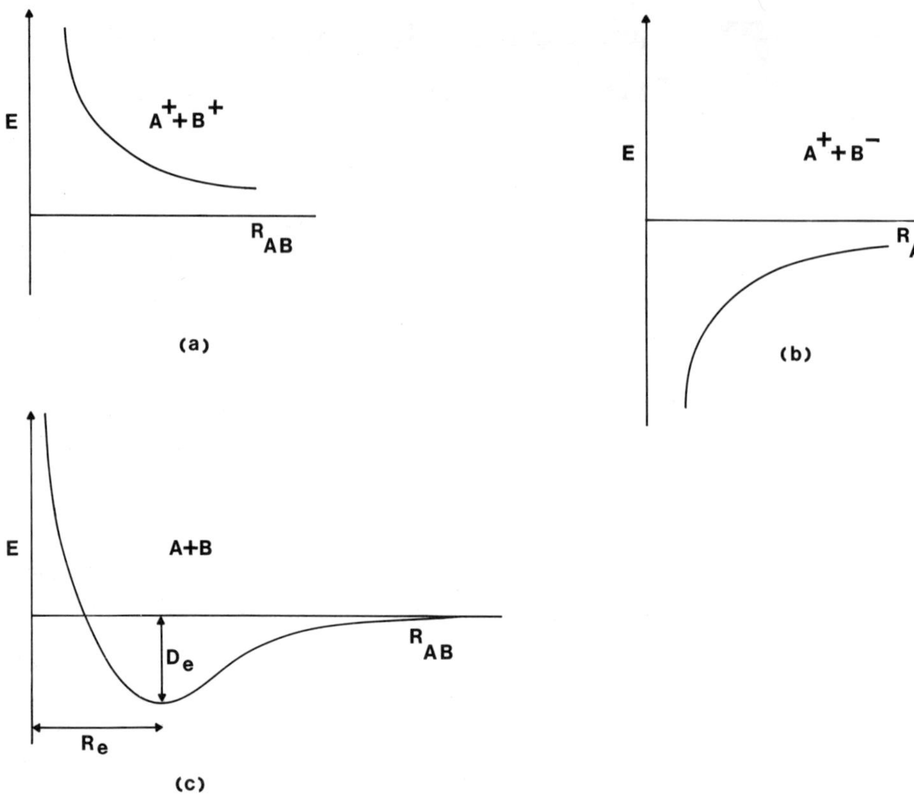

FIGURE 5.1. Variation of potential energy, E, with distance R_{AB} between atoms A and B.

is more stable when the two atoms are closer together. However, the potential energy increases very rapidly as the nuclei become closer than R_e. If no other factors intervene, the system will be most stable, and thus in an equilibrium situation, when the internuclear distance is R_e. This distance is termed the *equilibrium internuclear distance*.[3] Much of our discussion of valency is related to values of R_e for different atoms and also the corresponding lowering of the potential energy, D_e, that occurs when the two atoms come from a large distance apart to a separation of R_e. We see later that this represents the amount of energy required to break the chemical bond between A and B and so D_e is termed the *dissociation energy* of the bond.

[3]The equilibrium internuclear distance is frequently termed the *bond length* of the AB bond.

Let us now contemplate one of the fundamental questions of valency. Under what conditions will the elements A and B combine to form the chemical compound AB? To avoid unnecessary complications, we assume that the elements are in the form of an atomic gas (something that is realistic under ordinary conditions for the noble gases only) and also consider the compound AB as being in the form of gaseous diatomic molecules. Since the potential-energy story is told by Figure 5.1(c) it would seem that the compound AB will always form because it is more stable (i.e., has lower potential energy) than the separated A and B atoms. However, this neglects the fact that the atoms and molecules of the gas possess *kinetic energy* because of their thermal velocities. Some values of the average kinetic energy at different temperatures are given in Table 5.1 for a monatomic gas. For a diatomic gas the average energy is larger—at least 1⅔ as great—than for a monatomic gas at the same temperature because the diatomic molecules possess additional energy corresponding to rotational and vibrational motions.

So the potential energy of the diatomic molecules AB depicted in Figure 5.1(c) is only part of the story. The total energy of an AB molecule will be:

$$E_{\text{tot}} = \underset{\substack{\text{potential}\\ \text{energy}}}{V} + \underset{\substack{\text{kinetic}\\ \text{energy}}}{T}$$

and

$$T = \underset{\text{translation}}{T_{\text{trans}}} + \underset{\text{vibration}}{T_{\text{vib}}} + \underset{\text{rotation}}{T_{\text{rot}}}.$$

Since these kinetic energies are all positive, they tend to render the molecule less stable. Moreover, a molecule in a gas constantly undergoes collisions with other molecules (about 10^{10} s^{-1} at ordinary temperatures and pressures)

Table 5.1 Average kinetic energy per molecule of monatomic gas

T[K]	*E*[eV]	*E*[J]
100	0.0129	2.07×10^{-21}
200	0.0259	4.14×10^{-21}
273	0.0353	5.65×10^{-21}
300	0.0388	6.21×10^{-21}
1000	0.1293	2.07×10^{-20}

and energy tends to be shared among the molecules as a result of collision. Consequently, unless D_e, the dissociation energy of AB, is appreciably greater than the average thermal energy of gas molecules, any AB molecule that happens to be present will very quickly receive more kinetic energy from collisions than D_e, and the molecule will become less stable than the separated atoms—it will, in fact, come apart!

Hence we see that the necessary condition for forming a compound AB from the separate atoms A and B is that the lowering of the potential energy when A and B come together to the equilibrium internuclear distance must be substantially greater than the average thermal kinetic energy of the gas. If D_e is small we are not likely to detect the formation of AB except at low temperatures, whereas if D_e is large the molecule will be more easily produced and studied, even at much higher temperatures. For example, the molecule Ar_2 has $D_e = 0.0103$ eV $= 1.65 \times 10^{-21}$J and is difficult to detect, even at temperatures below room temperature, whereas the diatomic NaCl molecule can be observed in the vapor phase at temperatures in excess of 1000 K. It has $D_e = 4.24$ eV $= 0.68$ aJ.

We have confined our attention to diatomic molecules, but there are of course other possibilities to consider such as AB_2, A_2B, AB_3, and various large clusters containing various proportions of A and B. We must consider each possibility and determine whether the potential energy of the cluster of atoms is substantially lower than that of the corresponding numbers of separated atoms. The arrangement that has the lowest potential energy will be the favored, most stable one, but always the thermal kinetic energies are a competitive factor tending to disrupt a cluster. When we find the arrangement having the lowest potential energy we have predicted the *geometry of the molecule,* a property of great interest to chemists.

Valency and Quantum Mechanics

Because the constituents of a molecule—nuclei and electrons—are very tiny we must use quantum mechanics to calculate the properties of such systems. In principle we can solve all of the problems of valency, including the formation of molecules from atoms, their geometries, and other properties, by using quantum mechanics to compute the answers. In practice this is a formidable task, even for the simplest cases of two small atoms coming together to form a diatomic molecule. The arithmetic is so gigantic that even on the largest, fastest electronic digital computers we have to use some arithmetic simplifications to obtain answers and so the answers are only approximate.

But even if we could obtain exact answers it would not be a completely satisfactory theory of valency because a calculated value of D_e and R_e does not give us much understanding of the general kinds of chemical combination that we are likely to encounter throughout chemistry, or the broad trends in the geometries of various molecules and the kinds of chemical and physical properties that we may expect of the many compounds that we encounter in chemistry. We find it more instructive to develop approximate, pictorial (qualitative) approaches to guide us as to the rough magnitudes of dissociation energies, geometries, and the like. It is to these that we now turn as our guiding principles in the understanding of chemical compounds.

Combinations of Atomic Ions

Let us start with the situation most readily comprehended, one where simple electrostatics gives answers not very different from quantum-mechanical calculations—the combination of oppositely charged atomic ions A^+ and B^-. As was illustrated in Figure 5.1(b), the potential energy goes steadily down as the ions are brought together and we get the lowest value when the ions are touching. We can use the electrostatics formula

$$\Delta E = \frac{Q_1 Q_2}{4\pi\epsilon_0 R} \tag{5.1}$$

to calculate the change in potential energy, ΔE, as the ions are brought together from infinity. We need to know the charges Q_1 and Q_2 on A^+ and B^- and the internuclear distance for the touching ions, R. For singly charged ions we have

$$Q_1 = 1.602 \times 10^{-19}\ \text{C}\ (0.1602\ \text{aC})$$
$$Q_2 = -1.602 \times 10^{-19}\ \text{C}$$

(i.e., + and − the charge on a proton) and so

$$\Delta E = \frac{-(1.602 \times 10^{-19})^2}{(4)\,(3.142)\,(8.854 \times 10^{-12})\,(R)}\ \text{J}$$

$$\Delta E = \frac{-2.307 \times 10^{-28}}{R}\ \text{J}$$

where R is expressed in meters. It will be usual in molecular problems to express lengths in pm and for this case:

$$\left.\begin{aligned} \Delta E &= \frac{-230.7}{R}\ \text{aJ (R in pm)} \\ \text{or} \quad \Delta E &= \frac{-1440}{R}\ \text{eV (R in pm)} \end{aligned}\right\} \qquad (5.2)$$

If we are to apply this formula to help us decide whether two elements, such as Na and Cl, combine to form the compound NaCl, we must firstly remember that the elements are not initially charged, they exist as uncharged atoms (or atom clusters). We must, therefore, consider two processes: (a) conversion of the atoms to ions and (b) bringing the atomic ions together. Process (a) is a combination of two steps:

$$\begin{aligned} &\text{Na} \rightarrow \text{Na}^+ + \text{e}; \ \Delta E_1 = I(\text{Na}) = 5.14\ \text{eV} = 0.82\ \text{aJ} \\ &\text{Cl} + \text{e} \rightarrow \text{Cl}^- \quad \Delta E_2 = -A(\text{Cl}) = -3.62\ \text{eV} = -0.58\ \text{aJ}. \end{aligned}$$

Hence, for process (a):

$$\Delta E_a = \Delta E_1 + \Delta E_2 = +1.52\ \text{eV} = 0.24\ \text{aJ}.$$

The resultant ΔE_a is always positive; that is, the two separated ions are less stable than the separated atoms, no matter which two atoms we choose because the smallest known ionization potential of an atom (3.89 eV for Cs) is greater than the largest known electron affinity (3.62 eV for Cl).

The second process always results in a lowering of the potential energy. In other words, ΔE_b is always negative so that, provided that this lowering is sufficient to compensate for ΔE_a, the overall process corresponds to a lowering of potential energy and the two ions, close together, are more stable than the separated atoms. The various energy changes are illustrated in Figure 5.2(a).

It has been found by experiment that in the gaseous diatomic molecule NaCl the internuclear distance is 236 pm and so we can evaluate ΔE_b from Equation 5.2 as -0.977 aJ (or -6.10 eV). Hence for the overall production of the molecule from the atoms

$$\begin{aligned} \Delta E &= \Delta E_a + \Delta E_b \\ &= +1.52 - 6.10 \\ &= -4.62\ \text{eV}\ (-0.74\ \text{aJ}). \end{aligned}$$

This represents a considerably greater stability than necessary to withstand disruptive kinetic energies, even at temperatures around 1000 K and so offers a reasonable explanation of the stable existence of the sodium chloride molecule as a close combination of Na^+ and Cl^-. Notice

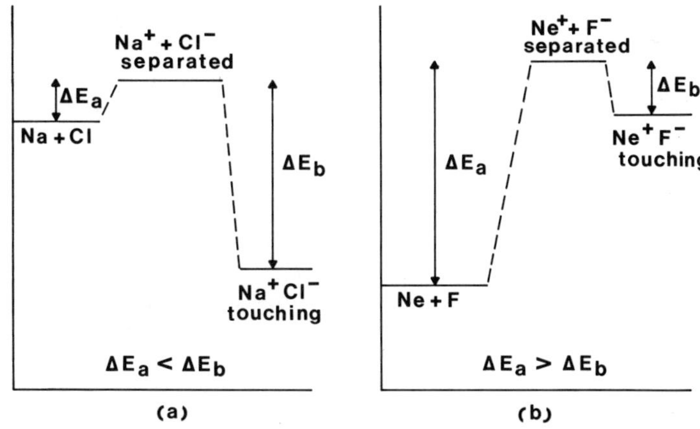

FIGURE 5.2. Energy changes for formation of Na^+Cl^- and Ne^+F^-.

that the alternative hypothesis of Cl^+ and Na^- is ruled out on stability grounds because in this case ΔE_a becomes 13.0 − 0.54 = 12.46 eV and so ΔE is 12.46 − 6.10 = +6.36 eV; hence the system is unstable. Likewise, the possibility Na^{++} and Cl^{--} can be eliminated. A similar explanation can be found for the stability of many other simple binary compounds such as KBr, CaO, and RbF.

In addition to explaining the existence and stability of many compounds, we can see that there are others for which the ionic bonding concept indicates that no stable combination is possible. Consider, for example, neon and fluorine

$$\left.\begin{array}{l} Ne \rightarrow Ne^+ + e \\ F + e \rightarrow F^- \end{array}\right\} \Delta E_a + 18.1 \text{ eV}.$$

In order to form a stable compound we must have ΔE_b at least as negative as −18.1 eV. Of course in principle we can achieve this lowering of energy by bringing the ions sufficiently close together. By rearranging Equation 5.2 we derive $R[\text{pm}] = 1440/\Delta E[\text{eV}]$; hence the Ne^+ and F^- ions need to approach to 1440/18.1 = 80 pm to achieve this energy decrease.

Such a close approach of these ions is not possible; the ionic radius of F^- alone is 136 pm (see Figure 4.15). Thus we can confidently predict that neon and fluorine do not enter into chemical combination as ions. They are more stable as separated atoms [see Figure 5.2(b)]. The ionic radii (Figure 4.15) based on experimental studies of the geometry of crystal lattices (see Chapter 6) are useful in deciding just how close various ions can approach and hence just how negative ΔE can be.

Ion Clusters

We know from chemical analysis that solid sodium chloride has the chemical composition NaCl and so in considering combinations of ions from sodium and chlorine atoms it was natural to take 1:1 ratios. But, as scientists, we should consider other alternatives. For example, what about Na^{2+} with two[4] Cl^-?

The comparison of relative stability that we have to make is: (a) one molecule of $NaCl_2$ plus a separate Na atom and (b) two molecules of NaCl. For a cation–anion internuclear distance of 236 pm in both cases and a linear Cl–Na–Cl arrangement for $NaCl_2$ we derive:

$$\text{for two NaCl:} \quad \Delta E = -9.2\ \text{eV}$$
$$\text{for NaCl}_2\ \text{plus Na:} \quad \Delta E = +23.8\ \text{eV}.$$

The much higher energy of $NaCl_2$ plus Na comes primarily from the very large second ionization potential of sodium (47.3 eV). This swamps the more favorable electrostatic energy of the $NaCl_2$ molecule.

Some analogous calculations for the combination of calcium and chlorine are outlined in greater detail in Appendix IV. Here the most favorable of the possibilities is $CaCl_2$. As a general rule, the cation that leads to the most favorable arrangement is the one obtained by *removing successive valence electrons until all are removed* (e.g., Na^+ and Ca^{2+}). Then the next ionization potential becomes far too large to make it worthwhile, energetically, to remove any further electrons. For anions there are insufficient data to study all possibilities, but it seems that the most stable arrangement is to *add electrons until the valence shell of the ion is completed* (e.g., Cl^- and O^{2-}).

Now although we have established that Na^+Cl^- is more stable than $Na^{2+}Cl^{2-}$ and is also preferable to $Na^{2+}(Cl^-)_2$, we have still not examined another possibility—one that turns out to be most important, namely, large clusters of Na^+ and Cl^-. A quite simple calculation shows that, just as Na^+Cl^- close together are more stable than separated Na^+ and Cl^-, so two Na^+ and two Cl^- arranged together in a square are more stable than two separate Na^+Cl^- molecules. And a tetramer $(Na^+)_4(Cl^-)_4$ in the form of a cube is

[4]Notice that we have to take a combination that is electrically neutral overall. It is not physically realistic to consider a large sample of, say, gaseous sodium chloride composed of Na^{2+} and Cl^- in a 1:1 ratio because, for example, one mole of such gas (ca. 58.5 g) would bear a total charge of $6.02 \times 10^{23} \times 1.60 \times 10^{-19} = 96$ kC. This is a fantastically large charge. For example, the field leading to dielectric breakdown and electric discharge in air at ordinary temperatures and pressures is around 3 MV m^{-1}. The field surrounding a charge of 96 kC is in excess of the air-breakdown figure up to a distance of about 20 km!

much more stable than four separate Na^+Cl^- (see Appendix IV for some calculations).

An enormously large, closely packed array of ions with Na^+ and Cl^- is even more stable in comparison with the equivalent number of diatomic molecules and turns out to be the most stable possible arrangement of the ions. Similarly, for other combinations of ions (e.g., CsBr, MgF_2, and BeO) a single, gigantic stack of closely packed ions turns out to be the most stable arrangement. The particular form of stacking that is most stable depends on the relative sizes of the ions and on the relative numbers of cations and anions. The specific arrangements are described in more detail in Chapter 6, where the structure of crystalline solids is considered. For each type of crystal it is possible, by a rather complex calculation, to derive a constant M, called the Madelung constant,[5] which tells us how much more stable is the very large stack of ions than the corresponding number of separate smallest molecules (diatomic, triatomic, etc., depending on the formula of the compound):

$$\Delta E_{\mathrm{b}} = \frac{N_{\mathrm{A}} M Q_1 Q_2}{4\pi\epsilon_0 R} \tag{5.3}$$

where N_{A} is Avogadro's number, R is the cation-to-nearest-anion distance, Q_1 and Q_2 are the charges on the cation and anion, and the result is the lattice energy per mole of compound. Table 5.2 lists some Madelung constants for different kinds of crystals.

[5] E. Madelung was a pioneer in calculations on ionic crystals. His work was published in 1909 and 1910.

Table 5.2 Madelung constants for some crystals

Type of Lattice	Composition[a]	M
Sodium chloride (NaCl)	1:1	1.7476
Cesium chloride (CsCl)	1:1	1.7627
Zinc blende (ZnS)	1:1	1.6381
Wurtzite (ZnS)	1:1	1.641
Fluorite (CaF_2)	1:2	5.0388
Cuprite (CuO_2)	1:2	4.1155
Rutile (TiO_2)	1:2	4.816
Anatase (TiO_2)	1:2	4.800
Corundum (Al_2O_3)	2:3	25.0312

[a] For 1 mole of separate simple molecules the potential energy is $N_A Q_1 Q_2/4\pi\epsilon_0 R$ for 1:1 composition, and 1.75 $N_A Q_1 Q_2/4\pi\epsilon_0 R$ for a 1:2 composition (and linear triatomic geometry).

Review of Ionic Bonding

In the preceding sections we have seen that it is possible to explain the existence and stability of a considerable number of chemical compounds by supposing that they consist of collections of oppositely charged ions. The most stable arrangements involve removing all of the valence electrons from one kind of atom to form a cation and of building up a complete shell of valence electrons on the other kind of atom to form an anion. Now because the substance has to be electrically neutral, the proportions in which the ions occur in the crystal are dictated by the charges on the respective ions; for example, Ca:Cl is 1:2 because the Ca^{2+} and Cl^- ions have charges in the ratio of 2:1. We thus arrive at an *octet rule* for the ions and valencies corresponding numerically to the periodic table group G or its "complement," 8 − G for the compound (e.g., Ca: G = 2 and Cl: 8 − G = 1 for $CaCl_2$). In regard to ions, the octet rule states:

> The most stable ionic solids come from atoms gaining or losing electrons to give ions with complete octets of outer electrons.

Notice that, while we have found a description that explains the existence of some compounds, this alone does not prove that the form of bonding is ionic attraction. There may be some other way by which the atoms can interact with one another that leads to even greater stability and so is preferred. However, it has been shown by very thorough quantum-mechanical studies that the combination of an electropositive element and an electronegative element normally involves oppositely charge ions. We also have careful experimental studies of the structure of crystals of compounds like sodium chloride (see Figure 5.3) that indicate the solid to be composed of ions.[6] The experimental estimate of the total number of electrons associated with the Na and Cl nuclei in sodium chloride is 9.98 and 17.85, respectively, close to the values of 10 for Na^+ and 18 for Cl^-.

Further evidence of the reality of ionic bonding is provided by measurements of electric dipole moments of molecules. These are discussed later (pp.122–124).

The Covalent Bond

Although we have found a credible explanation of the formation of a considerable number of chemical compounds by using the simple principles of electrostatics, it is

[6]The intensity of beams of X-rays scattered by a crystal lattice is determined by the electron density at different points in the crystal and so these intensities can be used to estimate electron densities in solids.

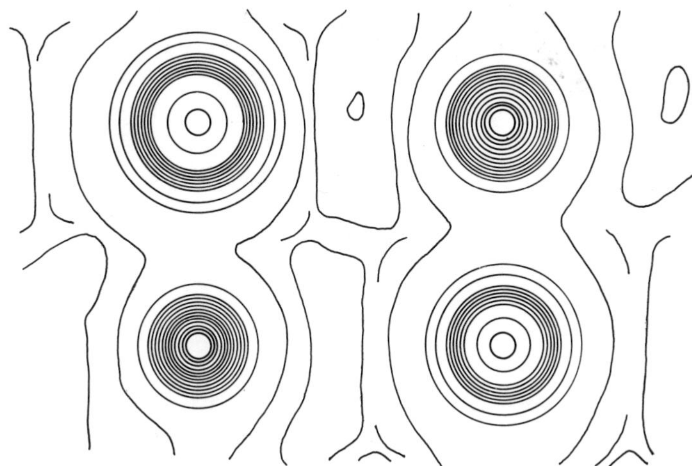

FIGURE 5.3. Electron-density contours for a section through a crystal of sodium chloride. Dense regions correspond to Cl^- ions, the less dense to Na^+ ions.

evident that this theory of ionic bonding does not explain the existence of compounds such as H_2, F_2, CH_4, NH_3, and C_2H_4. These compounds have very different properties from the ionic compounds just considered and to explain their stability we must seek some other mechanism by which the total energy can decrease as the atoms come together. An empirical explanation was offered first by G. N. Lewis in 1916 but a clear understanding did not emerge until quantum mechanics was applied to the problem in 1927 by W. Heitler and F. London. Let us start the discussion of this picture of chemical bonding by considering the simplest of all molecules—the hydrogen molecule–ion—H_2^+. We need to consider the change in energy that occurs when a proton, H^+, is brought up to a hydrogen atom.

Firstly, if the hydrogen atom consisted of a nucleus surrounded by a spherical electron cloud of finite size, say, of radius r, then to the outside proton the nucleus of the hydrogen atom is completely screened by the surrounding electron cloud and there is no resultant attraction or repulsion.[7] The change in energy as the proton approaches is then shown in Figure 5.4. It is only when the proton penetrates within the electron cloud that some change in

[7]Of course the field of the approaching proton would *polarize* the hydrogen atom (see pp. 122–124), but for the present discussion it will simplify our thinking if we assume that the hydrogen atom has a fixed, nonpolarizable structure.

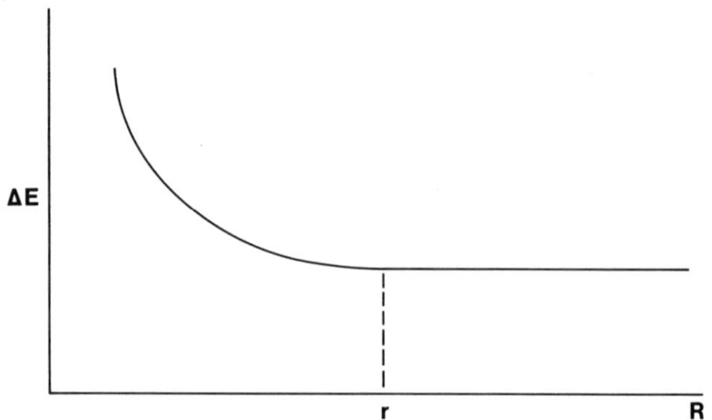

FIGURE 5.4. Proton approaching "hydrogen atom" of finite radius, *r*.

the electrostatic energy occurs—the nucleus is incompletely screened and the energy increases.

In Chapter 3 we saw that the electron density in a hydrogen atom (and other atoms) extends to infinity. In the case of the hydrogen atom the electron density diminishes exponentially with distance from the nucleus:

$$\psi_{1s}^2(\mathrm{H}) = \frac{1}{\pi a_0^3} \exp\left(\frac{-2r}{a_0}\right)$$

where r is the distance from the nucleus and a_0 is the Bohr radius. If we assume that the approaching proton does not disturb the electronic structure of the hydrogen atom, it is possible to evaluate the change in electrostatic energy as the proton moves in. The result is

FIGURE 5.5. Proton approaching unperturbed hydrogen atom.

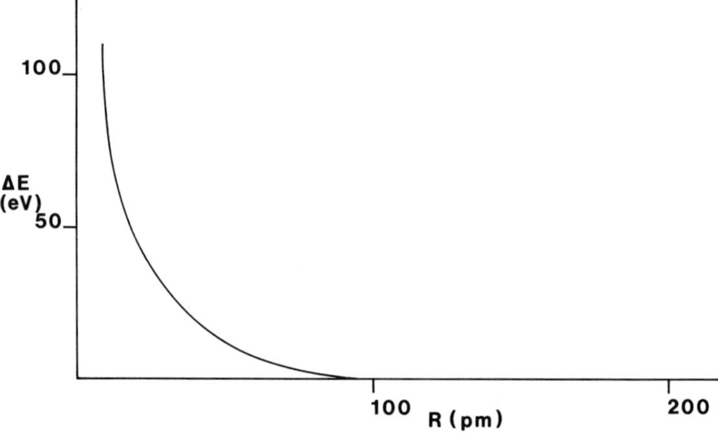

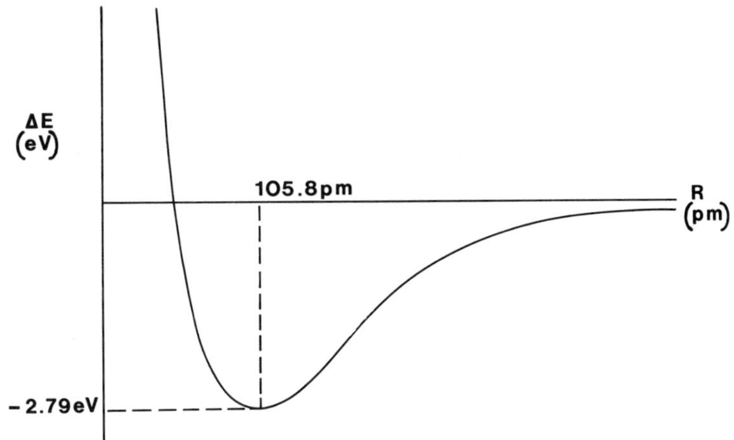

FIGURE 5.6. Energy versus internuclear distance for H_2^+.

$$\Delta E = 27.210 \exp\left(\frac{-2R}{a_0}\right)\left(1 + \frac{a_0}{R}\right) \text{ eV}.$$

The trend of ΔE with R is shown in Figure 5.5.

However, the variation of ΔE with R when a proton approaches a real hydrogen atom follows a very different pattern (see Figure 5.6). The reason is not too difficult to see. As soon as the proton comes close to the hydrogen atom, the electron in the latter is unable to distinguish between the two protons and so starts to share its time between the two. It starts to move in a way notably different from its former atomic state of motion described by the atomic orbital ψ_{1s}. Indeed, we say that its motion is now to be described by a *molecular orbital.* It turns out that in this state of motion the electron spends its time mostly in the region of space between the two nuclei and, as we shall see shortly, this has the effect of drawing the nuclei together and stabilizing the system. It is this new state of motion, represented by the molecular orbital, that leads to the minimum in the total energy curve of Fig. 5.6 and is fundamentally responsible for the stability of H_2^+ and many other diatomic and polyatomic molecules. For later use it will be convenient to think of the electron as shared by the nuclei.

The Virial Theorem and Molecular Stability

Now when we say that H_2^+ is a stable molecule we mean that its *total energy* is lower (by, in fact, 2.8 eV) than the total energy of a hydrogen atom and a separate proton. The total energy is the sum of the *kinetic energy of the electron* and the *potential energy* arising from the various electrostatic attractions and repulsions

$$\begin{aligned} E &= T + V \\ \text{total} &= \text{kinetic} + \text{potential}. \end{aligned} \tag{5.4}$$

We can use fairly simple arguments to judge the qualitative changes in potential energy of a system of electrons and nuclei but it is harder to picture trends in the kinetic energy of electrons.[8] It is thus very helpful to use a theorem, termed the *virial theorem,* which states that there is a simple relationship between kinetic, potential, and total energy when the atoms or molecules are in an equilibrium arrangement.

$$T = -\frac{V}{2} \tag{5.5}$$

hence

$$E = \frac{V}{2}. \tag{5.6}$$

We do not want to use this relationship quantitatively but merely to apply the qualitative principle:

> The change in total energy parallels the change in potential energy.

Thus if the total energy of a molecule is lower than that of its separated constituent atoms, it means that the potential energy is lower and vice versa. We can, therefore, focus our attention on the potential energy, which is easier to discuss, knowing that our conclusions apply also to the total energy. For a diatomic molecule, the region of space in which the potential energy of an electron is particularly low, and hence is a "favorable" region for contributing to molecular stability, is the internuclear region (shaded area in Figure 5.8). This lowering of potential energy by comparison with the situation in a separated atom results because the electron is simultaneously attracted strongly by two nuclei. The effect of this lowering of V has to be offset against the increase in potential energy coming from the repulsion between the nuclei. The electron attraction wins because the electron–nucleus distance is less than the nucleus–nucleus distance.

[8]From quantum mechanics it can be shown that the more confined the region of space in which an electron moves, the higher is its kinetic energy. But it is usually not very easy to judge qualitatively the "confinedness" of atomic or molecular orbitals since they always extend to infinity.

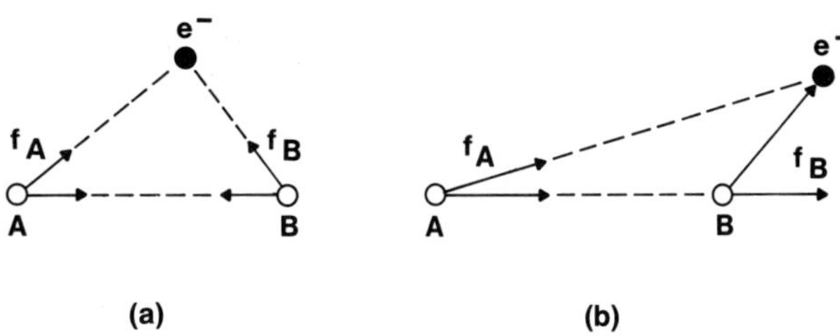

FIGURE 5.7. Forces in H_2^+.

The Forces in H_2^+

We can see this more clearly perhaps by considering the forces exerted on the nuclei by an electron (see Figure 5.7). We must pay particular attention to the components of the forces along the internuclear axis. Clearly in the case of Figure 5.7(a) the components result in *bringing the nuclei together,* whereas in Figure 5.7(b) the components have the overall effect of separating the nuclei (nucleus B is being pulled more to the right than nucleus A). By simple electrostatic calculation we can divide the space into two regions—*binding regions,* where the forces exerted by the electron draw the nuclei together and *antibinding regions,* where the forces push the nuclei apart. Figure 5.8 shows these regions for H_2^+.

The molecular orbital occupied by the electron in H_2^+ has a general shape such that the electron mostly stays in the binding region with the result that the electron acts somewhat like "nuclear cement" holding the nuclei near to each other in the form of a diatomic molecule.

The Molecular Orbitals 1sσ and 1sσ*

The specific mathematical expression for the molecular orbital may be derived (by elaborate mathematical procedures) by solving the Schroedinger equation for H_2^+, just as the expressions for atomic orbitals were derived by solving the equation for the hydrogen atom (see pp. 45–53). As in the case of the hydrogen atom, we actually obtain a

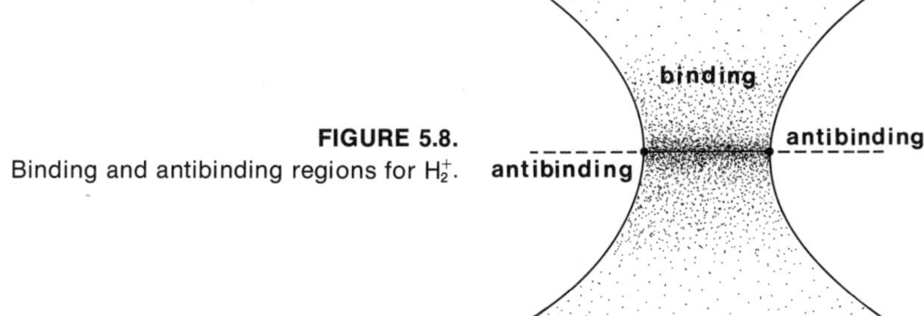

FIGURE 5.8.
Binding and antibinding regions for H_2^+.

whole family of solutions—various molecular orbitals—representing the various permitted states of motion of the electron in H_2^+. The particular molecular orbital that we were discussing in previous sections is actually the one of lowest possible energy, akin to the ls orbital of the hydrogen atom. For reasons discussed below, it is called the "$1s\sigma$" molecular orbital. The other molecular orbitals, of higher energy, represent states of motion of the electron that do not lead to sufficient molecular stability because the electron is not as much restricted to the binding region of space as it is in the $1s\sigma$ orbital.

The higher-energy molecular orbitals are of great importance for helping us to understand the electronic structures of more complex diatomic molecules, and we use them in much the same way as we used the scheme of atomic orbitals to understand atomic structure.

To help us to draw up the scheme of molecular orbitals it is useful to picture what happens to the atomic orbitals when two atomic nuclei, initially infinitely separated, are steadily brought together. The atomic orbitals smoothly change into the corresponding molecular orbitals. For example, the two ls atomic orbitals change into two molecular orbitals, one having a lower energy than the original atomic orbitals, the other higher as shown by the following schematic diagram (Figure 5.9).

When electrons occupy the lower-energy molecular orbital they tend to make the molecule more stable than the separated atoms and so the molecular orbital is called a *bonding molecular orbital.* Electrons in the higher-energy molecular orbital would tend to decrease the stability of the molecule and so the molecular orbital is called an *antibonding molecular orbital.*

FIGURE 5.9. Energy-level diagram showing molecular-orbital (MO) formation.

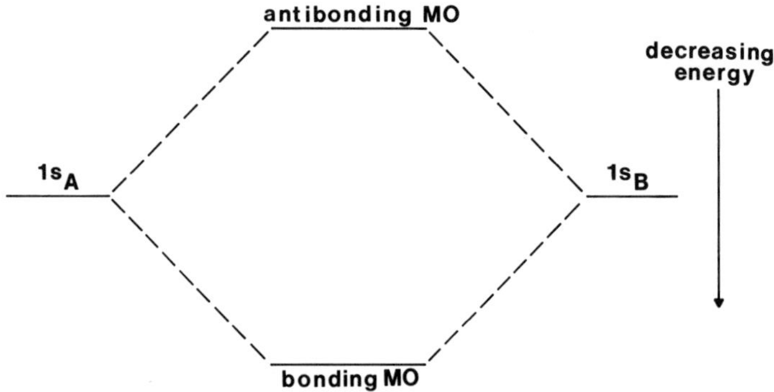

In contrast to the bonding orbital, the antibonding orbital confines the electron mainly to the antibinding region of space. The relationship between the bonding–antibonding nature of the molecular orbitals and the binding–antibinding parts of space can be seen by comparing Figure 5.10, which shows the derivation of the molecular orbitals from atomic orbitals, with Figure 5.8.[9]

If we view these orbitals along a line passing through the two nuclei, from this direction the functions are cylindrically symmetrical. This is the molecular counterpart of the spherical symmetry of s orbitals and so these molecular orbitals are termed σ-orbitals. It is customary to indicate antibonding orbitals by an asterisk and so the two molecular orbitals illustrated in Figures 5.9 and 5.10 are usually denoted by 1sσ and 1sσ*, with the "1s" to remind us which orbitals of the separated atoms are related to the molecular orbitals. Other molecular orbitals that we encounter below do not have cylindrical symmetry about the internuclear axis but look like a side view of a p-type atomic orbital when viewed along the internuclear direction. They are termed π-orbitals. The correspondence can be extended to δ- and ϕ- molecular orbitals that relate to d- and f-type atomic orbitals:

Atomic orbitals	s p d f . . .
Molecular orbitals	σ π δ ϕ . . .

[9]The question of "where the electrons are" of course can only be answered by the probability of finding them in a given region of space. This probability is proportional to the square of the atomic and molecular orbitals, which will have a shape similar to those given for the orbitals in Figure 5.10 for say, the 90% probability of finding the electrons in the regions shown.

FIGURE 5.10. Pictorial representation of molecular-orbital formation.

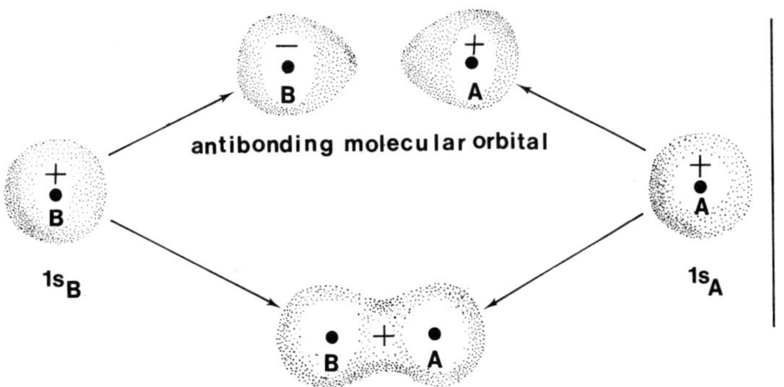

Greek letters correspond to molecular orbitals and Roman letters to atomic orbitals. However, we need to consider only σ- and π-type molecular orbitals.

It is helpful to picture the molecular orbitals as sums and differences of the related atomic orbitals.

$$1s\sigma = 1s_A + 1s_B \tag{5.7}$$
$$1s\sigma^* = 1s_A - 1s_B \tag{5.8}$$

(notice how the two combinations account for the relative signs of the parts of orbitals as illustrated in Figure 5.10).

This approximate representation of molecular orbitals is not too bad mathematically, as can be seen in Figure 5.11, where the exact functions for $1s\sigma$ and for $1s\sigma^*$ are compared with the approximate functions given in equations (5.7) and (5.8) [such approximations are referred to as linear combinations of atomic orbitals (LCAO)].

The Molecules H_2^+, H_2, He_2^+, and He_2

Each molecular orbital, whether it is a bonding molecular orbital or an antibonding molecular orbital, can accommodate only two electrons. Again, this is a consequence of the Pauli principle.

Furthermore, the building up of the electron configurations of diatomic molecules is exactly the same as we saw in Chapter 4 for building up the electron configurations of atoms. We first establish a "filling order," then we use the Aufbau principle, the Pauli principle, and Hund's rule in order to write down the electron configuration.

As was pointed out in the preceding paragraphs, the $1s\sigma$ is the lowest-energy molecular orbital. The $1s\sigma^*$ molecular orbital corresponds to the next lowest-energy solution of the Schroedinger equation for H_2^+; that is, it is the next lowest-energy molecular orbital. We note that these two molecular orbitals of lowest energy are formed from linear combinations of the lowest-energy atomic orbitals. These

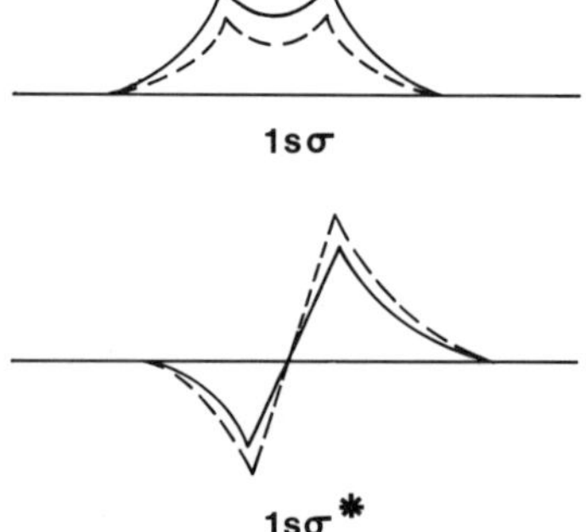

FIGURE 5.11. Comparison of exact functions for $1s\sigma$ and $1s\sigma^*$ with the linear combination of atomic orbitals (LCAO) approximation. Dashed lines are the LCAO approximated molecular orbitals.

two molecular orbitals can accommodate a total of four electrons (two in each orbital), and so let us consider in detail some simple homonuclear diatomic molecules with up to four electrons.

H_2^+

The hydrogen-molecule ion has but one electron, and so this goes into the $1s\sigma$ bonding molecular orbital. The configuration is written

$$H_2^+ \ (1s\sigma)^1.$$

H_2

The hydrogen molecule has a total of two electrons, both of which go into the $1s\sigma$ molecular orbital, hence the configuration is written

$$H_2 \ (1s\sigma)^2.$$

He_2^+

The helium molecule ion has three electrons. The first two are accommodated in the $1s\sigma$ bonding MO and the third, in the $1s\sigma^*$ antibonding molecular orbital.

$$He_2^+ \ (1s\sigma)^2(1s\sigma^*)^1.$$

He_2

For the helium molecule with four electrons the $1s\sigma$ and $1s\sigma^*$ molecular orbitals are filled, each with two electrons

$$He_2 \ (1s\sigma)^2(1s\sigma^*)^2.$$

The electron configurations we have just written immediately suggest several ideas that are amenable to experimental confirmation. First of all, is the concept of *net bonding,* that is, whether it is even possible to form the molecule. We again turn to our criterion that a molecule is stable with respect to its constituents if the energy is lowered. Clearly this is the case for the $1s\sigma$ bonding molecular orbital, as can be seen from Figure 5.9 or 5.10. However, the $1s\sigma^*$ orbital results in a raising of the energy by at least an equivalent amount. We define the *net bonding electrons* as the number of electrons in bonding molecular orbitals minus the number of electrons in antibonding molecular orbitals. For bond formation, the number of net bonding electrons must be greater than zero. Thus for our simple molecules, we expect H_2^+, and H_2, and He_2^+ to be stable with respect to their constitutent atoms, i.e. a bond to be formed. We expect He_2 to be unstable, with no strong bond formed. These expectations are confirmed experimentally in that H_2^+, H_2, and He_2^+ are all well-known species, whereas He_2 is only very weakly associated (see p. 164). Chemists often find it useful to use a term called the *bond order* (BO), which is defined as:

$$\mathrm{BO} = \frac{1}{2}\{\text{net bonding electrons}\}.$$

Thus we can construct a table:

Molecule	Net Bonding Electrons	Bond Order	Common Name
H_2^+	1	½	half-bond
H_2	2	1	single-bond
He_2^+	1	½	half-bond
He_2	0	0	—

Similarly, a bond order of two is called a double bond, and a bond order of three is called a triple bond. We encounter these higher bond orders later.

Secondly, we should see a correlation between the experimentally observed bond length and our molecular-orbital description. Here the more net bonding electrons there are, the more tightly the two nuclei should be drawn together and hence the internuclear distance should be shorter. This turns out to be the case because the experimental values are:

Molecule	Bond Length
H_2^+	106 pm
H_2	74 pm
He_2^+	108 pm

Thirdly, we expect a correlation between the dissociation energy and our MO description. The experimental numbers

Molecule	D_e [eV]
H_2^+	2.65
H_2	4.48
He_2^+	3.10

confirm our expectations.

Finally, molecules with unpaired electron spin (i.e., unfilled orbitals) have the property that they are strongly attracted by an external magnetic field. These molecules are termed *paramagnetic,* and we confidently predict with our molecular orbital picture that He_2^+ and H_2^+ will exhibit this behavior. On the other hand, molecules in which all of the electrons have paired spins are weakly repelled by a magnetic field. These molecules are termed *diamagnetic,* an example being H_2. The experimental magnetic properties of H_2^+, H_2, and He_2^+ are all in accord with these theoretical predictions.

The Molecular Orbitals $2s\sigma$, $2s\sigma^*$, $2p_z\sigma$, $2p_z\sigma^*$, $2p_x\pi$, $2p_x\pi^*$, $2p_y\pi$, and $2p_y\pi^*$

For diatomic molecules with more than four electrons, further molecular orbitals can be constructed by taking appropriate linear combinations of atomic orbitals.

Consider first the molecular orbitals formed by bringing together two 2s orbitals. Again we see that two molecular orbitals are formed, one having an energy lower than the original atomic orbitals and called the "$2s\sigma$" bonding molecular orbital, the other having higher energy and called the "$2s\sigma^*$" antibonding molecular orbital. Schematically the process is shown in Figure 5.12. As with the $1s\sigma$ molecular orbital, the $2s\sigma$ molecular orbital concentrates electron density between the two nuclei, whereas in the $2s\sigma^*$ molecular orbital there is a nodal plane of zero electron density between the nuclei.

Because the 2s atomic orbitals are of higher energy that the 1s atomic orbitals, we expect the $2s\sigma$ and $2s\sigma^*$ molec-

FIGURE 5.12. The formation of the $2s\sigma$ and $2s\sigma^*$ molecular orbitals from two 2s atomic orbitals.

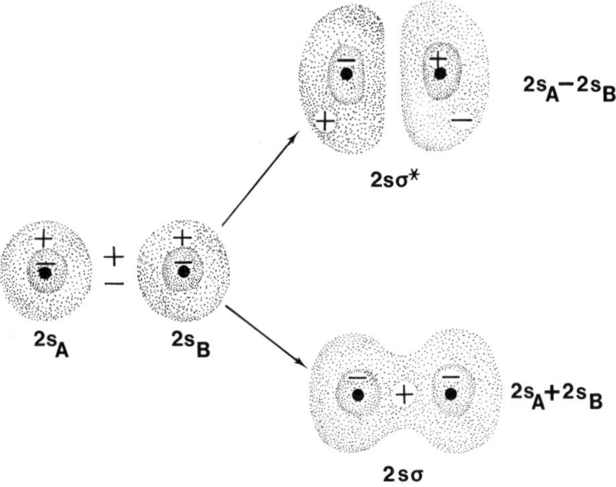

ular orbitals to be of higher energy than the $1s\sigma$ and $1s\sigma^*$ molecular orbitals. This is indeed the case and so our filling order to date reads:

$$1s\sigma < 1s\sigma^* < 2s\sigma < 2s\sigma^*$$

and we can accommodate up to eight electrons in these four molecular orbitals.

Next we examine the molecular orbitals that can be constructed from the three 2p atomic orbitals. Here there are two cases we must consider because the p-orbitals have directional properties. We usually define our diatomic molecule such that the z-axis corresponds to the line connecting the two nuclei. Thus as two $2p_z$ atomic orbitals are brought together to form molecular orbitals they approach each other head-on (see Figure 5.13). The molecular orbitals are once again cylindrically symmetric and thus are of the σ-type. They are called the $2p_z\sigma$ *bonding molecular orbital* and $2p_z\sigma^*$ *antibonding molecular orbital.*

The remaining 2p orbitals, the $2p_x$ and $2p_y$ on two atoms brought together, must approach each other in a parallel fashion. The molecular orbitals formed in this case do not have cylindrical symmetry and are called *π-molecular orbitals* (see Figure 5.14). The designation of the π-molecular orbitals follows that for the σ-molecular orbitals. If the molecular orbital concentrates electron density between the nuclei it is called a $2p_x\pi$ or $2p_y\pi$ bonding molecular orbital; if not, it is called a $2p_x\pi^*$ molecular orbital or $2p_y\pi^*$ molecular orbital. We might expect from the similar way in which they were formed and the fact that the 2p atomic orbitals are degenerate, that the $2p_x\pi$ and $2p_y\pi$ molecular orbital would be degenerate in energy as would the $2p_x\pi^*$ and $2p_y\pi^*$. This is indeed the case.

FIGURE 5.13. Molecular orbitals $2p_z\sigma$ and $2p_z\sigma^*$ formed from two $2p_z$ atomic orbitals.

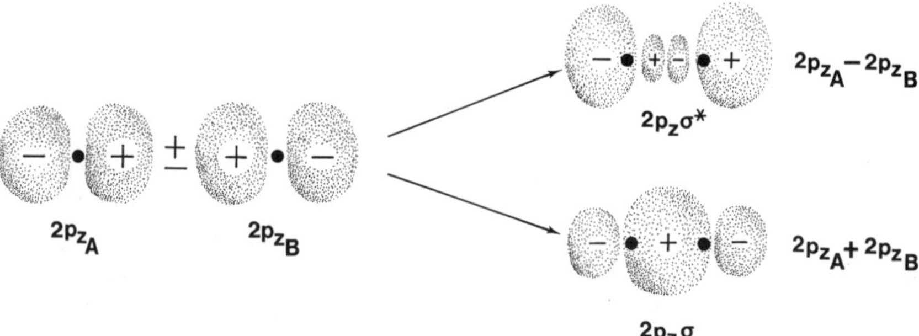

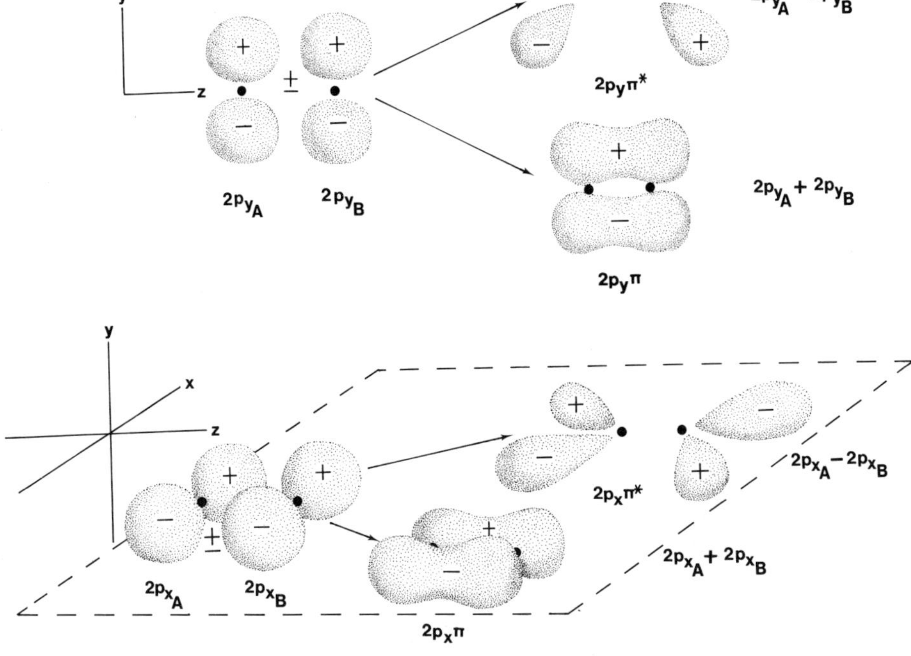

FIGURE 5.14. Molecular orbitals $2p_x\pi$, $2p_y\pi$ $2p_x\pi^*$, and $2p_y\pi^*$ formed from $2p_x$ and $2p_y$ atomic orbitals.

We now have a total of 10 molecular orbitals and hence should be able to describe properly the electronic structure of homonuclear diatomic molecules containing up to 20 electrons.

One small hitch is that we must know the "filling order." This is obtained from experiment and more complete calculations. The correct "filling order" for most homonuclear diatomic molecules is:

$$1s\sigma < 1s\sigma^* < 2s\sigma < 2s\sigma^* < 2p_x\pi = 2p_y\pi < 2p_z\sigma$$
$$< 2p_x\pi^* = 2p_y\pi^* < 2p_z\sigma^*.$$

This orbital energy scheme is depicted diagramatically in Figure 5.15(a). Because the $2p_z\sigma$ and $2p_x\pi = 2p_y\pi$ molecular orbitals are close to each other in energy, this "filling order" is not rigorous and a crossover in filling order is observed experimentally for O_2 and F_2. This alternative "filling order" is shown in Figure 5.15(b).

It is instructive to run through some first-row homonuclear diatomic molecules to illustrate what we have just learned. We have done this already for the molecules H_2^+,

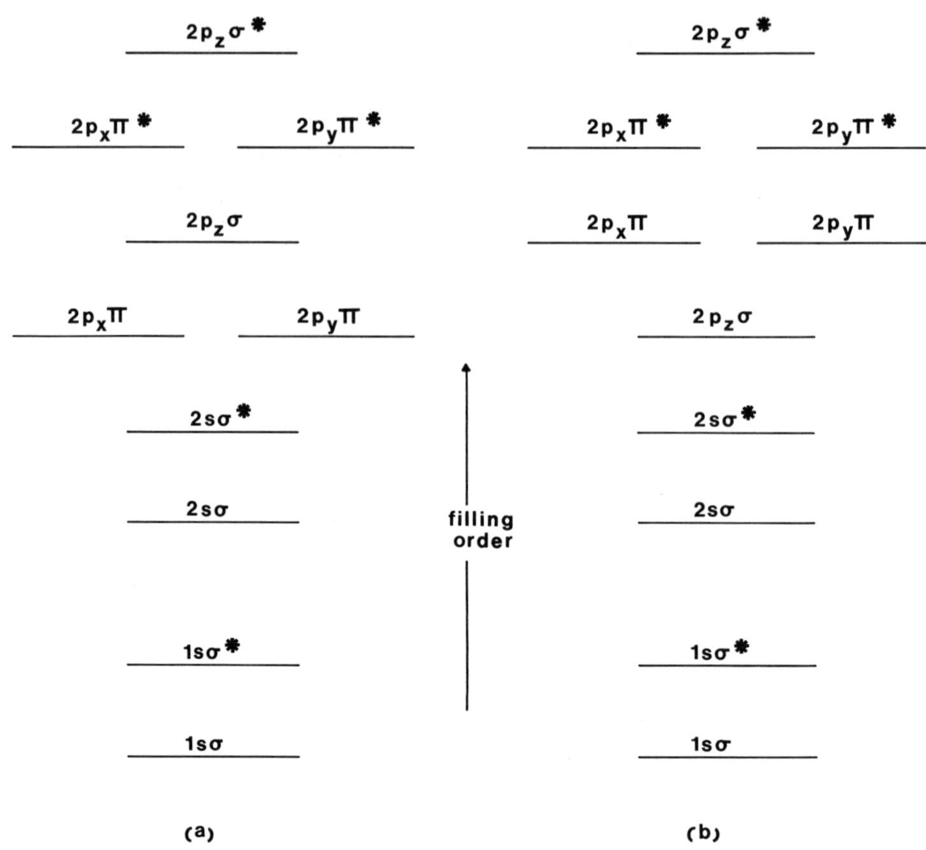

FIGURE 5.15. Homonuclear diatomic-molecule molecular-orbital filling order: (a) for all first-row homonuclear diatomics except O_2 and F_2; (b) for O_2 and F_2. Note crossover of the filling order of $2p_z\sigma$ and $2p_x\pi = 2p_y\pi$.

H_2, He_2^+, and He_2 involving just the $1s\sigma$ and $1s\sigma^*$ molecular orbitals. Moving now to the second row of the periodic table we begin with the diatomic molecule Li_2.

Li_2

Each lithium atom has three electrons, making a total of six electrons in Li_2. The configuration is written from the filling-order diagram Figure 5.15(a).

$$Li_2\text{: } (1s\sigma)^2(1s\sigma^*)^2(2s\sigma)^2$$

There are two net bonding electrons in the $2s\sigma$ molecular orbital; hence Li_2 has a single bond. The bond length of Li_2 is 267 pm, compared with 74 pm in H_2. This is because the average radius of the 2s orbitals forming the molecular orbital in Li_2 is greater than the average radius of the 1s orbitals forming the molecular orbital in H_2. In fact, the 1s atomic orbitals in Li_2 never get close enough to each other to form effective molecular orbitals. Probably a more realistic way of writing the configuration for Li_2 would be

$$Li_2\text{: } (1s_A)^2(1s_B)^2(2s\sigma)^2$$

but the net result is the same as the configuration obtained from the filling-order diagram. That is, the $1s\sigma$ and $1s\sigma^*$ molecular orbitals are actually little more than doubly occupied 1s atomic orbitals, one on each Li atom in the molecule.

Be_2

Here with eight electrons we arrive at the configuration

$$Be_2\text{: } (1s\sigma)^2(1s\sigma^*)^2(2s\sigma)^2(2s\sigma^*)^2.$$

There is no net bonding; therefore, we do not expect Be_2 to exist as a stable molecule. This is entirely consistent with experimental observation.

B_2

Diatomic boron gives the configuration from the filling-order diagram, Figure 5.15(a)

$$B_2\text{: } (1s\sigma)^2(1s\sigma^*)^2(2s\sigma)^2(2s\sigma^*)^2(2p_x\pi)^1(2p_y\pi)^1.$$

Notice that we have invoked Hund's rule in that the seventh and eighth electrons are put into separate degenerate π-molecular orbitals with parallel spins. Consequently we predict paramagnetic behavior for B_2, and this is observed experimentally.

As in the case of Li_2, the 1s atomic orbitals on the two nuclei do not come close enough to be significantly changed and it is probably better to think of the inner shell configuration as $(1s_A)^2$ $(1s_B)^2$ rather than $(1s\sigma)^2$ $(1s\sigma^*)^2$. The same remarks apply, although we do not repeat them here, to C_2, N_2, O_2, and F_2.

C_2

The configuration for diatomic carbon with 12 electrons is:

$$C_2\text{: } (1s\sigma)^2(1s\sigma^*)^2(2s\sigma)^2(2s\sigma^*)^2(2p_x\pi)^2(2p_y\pi)^2.$$

There are four net bonding electrons; hence the bond order is 2 and we say that C_2 has a double bond. The dissociation energy or so-called bond energy of C_2 is 6.5 eV molecule^{-1}. Comparing this with the bond energy of

3.0 eV molecule^{-1} for B_2 confirms that double bonds concentrate more electrons between the two nuclei than single bonds and are thus stronger bonds.

N_2

Moving on to diatomic nitrogen, the configuration is:

$$N_2\colon (1s\sigma)^2(1s\sigma^*)^2(2s\sigma)^2(2s\sigma^*)^2(2p\pi)^4(2p_z\sigma)^2.$$

Notice that we have grouped the $2p_x\pi$ and $2p_y\pi$ orbitals together here because they are degenerate. The bond order for N_2 is 3, and the bond energy of 9.8 eV molecule^{-1} accordingly indicates a very strong bond.

O_2

For the 16-electron oxygen molecule the filling-order diagram, Figure 5.15(b), is used giving the configuration:

$$O_2\colon (1s\sigma)^2(1s\sigma^*)^2(2s\sigma)^2(2s\sigma^*)^2$$
$$(2p_z\sigma)^2\ (2p\pi)^4(2p_x\pi^*)^1(2p_y\pi^*)^1.$$

The oxygen bond order is two and the molecule is paramagnetic, exactly as predicted by the configuration, because of the two unpaired electrons. The bond energy of 5.1 eV molecule^{-1} is consistent with a double bond.

F_2

The configuration for diatomic fluorine follows:

$$F_2\colon (KK)^4(2s\sigma)^2(2s\sigma^*)^2(2p_z\sigma)^2(2p\pi)^4 2p\pi^*)^4.$$

Here the designation KK indicates the molecular orbitals formed from the atomic orbitals of the K shell—that is, the $1s\sigma$ and $1s\sigma^*$ molecular orbitals or $(1s_A)^2\ (1s_B)^2$. The bond order of fluorine is unity. The bond energy is 1.6 eV molecule^{-1}.

Ne_2

The diatomic neon molecule would fill the 10 molecular orbitals that we have generated to date so that it must have zero net bonding electrons and not exist under ordinary conditions.

Diatomic Molecular Ions

It is a simple matter to add or subtract electrons to our diatomic molecules in order to form negative and positive ions. From the configuration that results we can predict the same kind of properties that were predicted for the

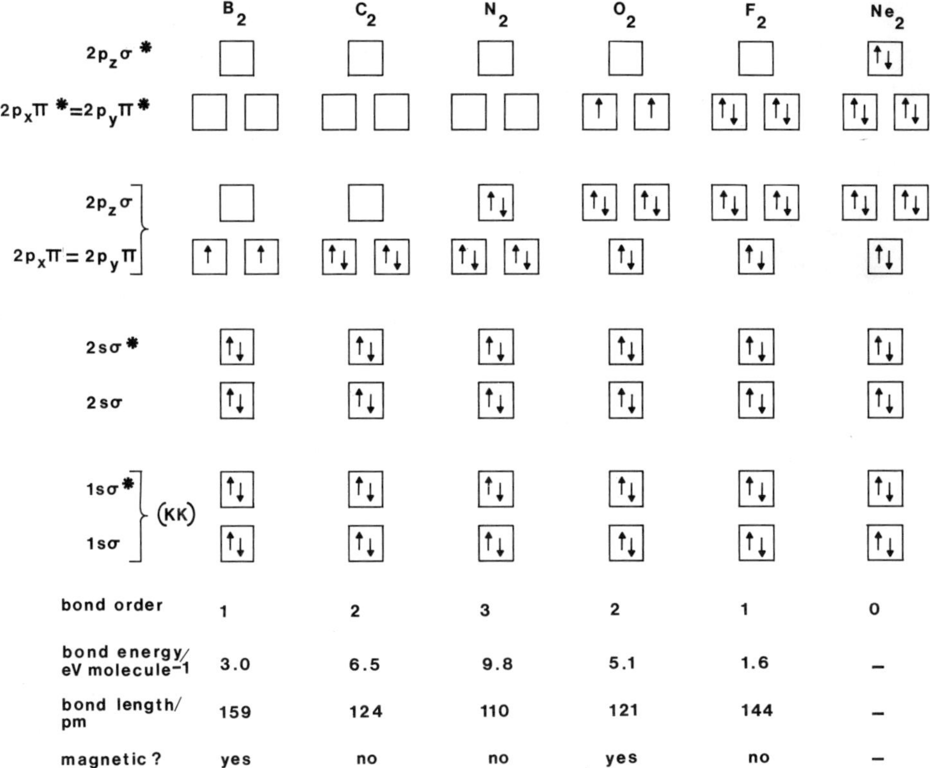

FIGURE 5.16. Molecular orbitals and properties of first-row homonuclear diatomic molecules.

neutral molecules. For example, forming negative ions of N_2, we find:

$$N_2^-\text{: } (KK)^4(2s\sigma)^2(2s\sigma^*)^2(2p\pi)^4(2p_z\sigma)^2(2p\pi^*)^1$$
$$N_2^{2-}\text{: } (KK)^4(2s\sigma)^2(2s\sigma^*)^2(2p\pi)^4(2p_z\sigma)^2(2p_x\pi^*)^1(2p_y\pi^*)^1.$$

Both of these ions ought to be stable because they have bond orders of 2½ and 2, respectively. Both should also be paramagnetic since there are unpaired electrons present. From what limited experimental evidence there is on these molecules these expectations are confirmed.

We summarize the properties, both predicted and experimental, for first-row homonuclear diatomics in Figure 5.16.

Heteronuclear Diatomic Molecules

Use of the molecular-orbital concept to explain the bonding characteristics of homonuclear diatomic molecules can, with a little care, be extended to deal with heteronuclear diatomic molecules such as LiH, HF, CO, and NO.

These four molecules will nicely illustrate how this can be done, but first we need to define our ground rules a bit more carefully. Namely, which atomic orbitals on the separate atoms can we combine to form the molecular orbitals? There are two guiding principles, specifically, that the atomic orbitals should have: (a) similar energies and (b) proper symmetry, so that the resulting bonding molecular orbital represents a lower energy state than the isolated atomic orbitals.

In principle (a) we refer to the atomic orbital energies in Figure 4.9 to decide which combinations of atomic orbitals are most likely for a given pair of atoms. For example, the orbital energies of Li, from Figure 4.9, are:

$$\begin{array}{ll} 2s & -5.4 \text{ eV} \\ 1s & -65 \text{ eV}. \end{array}$$

Since the hydrogen atom 1s orbital energy is −13.6 eV, the combinations of atomic orbitals appropriate to form the molecular orbitals is:

$$1s_H \pm 2s_{Li}.$$

Principle (b) merely states that the bonding molecular orbital formed must concentrate electrons between the two nuclei. The dissimilar atomic orbitals capable of doing this that we consider here are as follows for the molecule AB:

1s orbital on A ± 2s orbital on B, and vice versa
1s orbital on A ± $2p_z$ orbital on B, and vice versa
2s orbital on A ± $2p_z$ orbital on B, and vice versa

where we pick z as the internuclear axis. All of these molecular orbitals are of the σ-type (i.e., they have cylindrical symmetry).

A consequence of there being some difference in energies of the two atomic orbitals forming the molecular orbital is that the electrons in the molecular orbital will no longer be equally shared by the two nuclei. That is, they will spend on the average more time close to one nucleus than the other. The term that describes this ability of an atom to attract electrons of a chemical bond to itself is *electronegativity*. The scale of electronegativities that is perhaps most rigorously justified is that due to R. A. Mulliken, who represented the electronegativity by the mean of the ionization potential and electron affinity of an atom:

$$\chi = \frac{1}{2}(I + A) \tag{5.9}$$

A more widely used scale of electronegativity, χ, is that due to Linus Pauling, who defined it in terms of the bond energies of the molecules A_2, B_2, and AB. His argument is that, if the electrons in the bond of the molecule AB are equally shared, the AB bond energy should be simply the geometric mean[10] of the A_2 and B_2 molecules. Any deviation from the mean value is an indication that the electrons are not equally shared.

Pauling's formula for electronegativities is

$$\chi_A - \chi_B = D_e(AB) - [D_e(A_2)D_e(B_2)]^{1/2} \tag{5.10}$$

where χ_A represents the electronegativity (in eV) of atom A and $D_e(AB)$, $D_e(A_2)$ and $D_e(B_2)$ are dissociation energies (bond energies) of the molecules AB, A_2, and B_2, respectively. By defining the electronegativity of one element of the periodic table, all other electronegativities can be obtained relatively. Pauling chose the value for fluorine, the most electronegative element, as 3.98 eV. Other electronegativities on this basis are given in Table 5.3.

The LiH Molecule

The 2s orbital on Li with an orbital energy of −5.4 eV can form two molecular orbitals with the 1s orbital of H with an orbital energy of −13.6 eV, as shown in Figure 5.17.

[10]There are two ways to average two numbers X and Y:

$$\text{Arithmetic mean} = \frac{X + Y}{2}$$

$$\text{Geometric mean} = \sqrt{XY}$$

Pauling showed that we want the geometric mean.

Table 5.3 Pauling scale of electronegativities of some elements

Atom	χ [eV]	Atom	χ[eV]
H	2.20	Na	0.93
Li	0.98	Si	1.90
Be	1.57	P	2.19
B	2.04	S	2.58
C	2.55	Cl	3.16
N	3.04	Br	2.96
O	3.44	I	2.66
F	3.98		

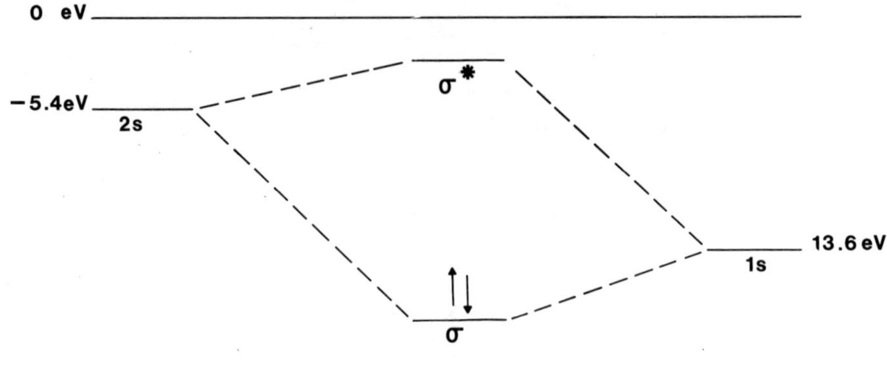

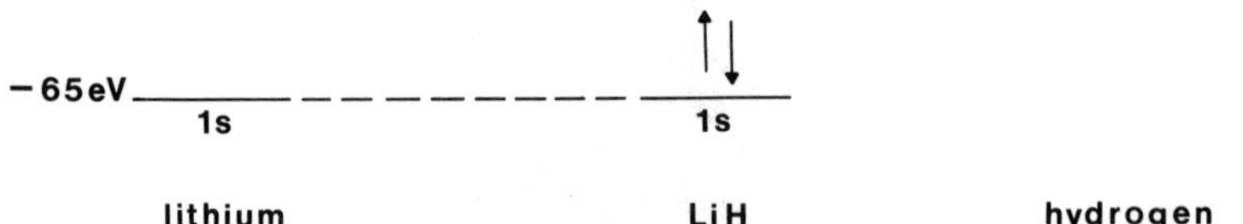

FIGURE 5.17. Molecular orbitals of LiH. The 1s orbital on Li remains essentially unchanged on the Li nucleus in LiH. The Li 2s and H 1s orbitals from a σ-bonding molecular orbital and a σ^* antibonding molecular orbital.

Notice that the Li 1s orbital is of too low an energy to combine effectively with the H 1s orbital. It remains essentially unaltered around the Li nucleus in the LiH molecule. The four electrons—three from Li and one from H—are then fed into the available orbitals, according to the Aufbau and the Pauli principles, the first two into the 1s orbital and the next two into the σ bonding molecular orbital. The bond order is thus equal to 1.

Hydrogen is more electronegative than Li, and so we expect the two electrons in the σ-molecular orbital to spend proportionately more time near the hydrogen nucleus. Thus there should be some separation of charge, which we could represent as:

$$\overset{\delta+}{\mathrm{Li}} - \overset{\delta-}{\mathrm{H}}$$

the δ implying some fractional part of a unit charge.

Alternatively, the σ-molecular orbital itself could be represented as skewed towards the H nucleus in a manner such as:

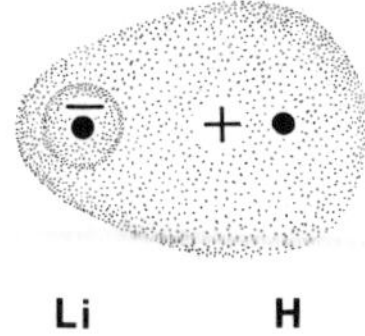

The separation of charge due to skewed electron distribution in a molecule means that the molecule behaves as a tiny *electrostatic dipole*. This dipolar property of molecules is an important property by which we can check the accuracy of our bonding theories. We discuss it further in a moment.

The skewedness of the charge distribution in a particular bond is often termed *ionic character*. As we approach 100-percent ionic character the diatomic molecule is described in terms of our electrostatic theory of ionic bonding. At the other end of the scale, diatomic molecules with little or no ionic character are more conveniently described by molecular-orbital theory. It is important to remember that both descriptions are only approximate and are used solely because they give theoretical answers that are close to experimentally observed properties of the diatomic molecule.

The HF Molecule

The fluorine orbital energies are:

1s	−697 eV
2s	−40.1 eV
2p	−18.6 eV.

Obviously from our energy criterion only the 2p orbitals have an orbital energy close enough to combine with the H 1s orbital energy of −13.6 eV. To form molecular orbitals with the p-orbitals, only the $2p_z$ orbital has the correct directional properties to form any effective molecular orbitals. We can write the electron configuration for HF as:

$$\text{HF: } (1s)^2(2s)^2(2p_x)^2(2p_y)^2\sigma^2.$$

Here the σ-molecule orbitals have a shape something like that depicted in Figure 5.18.

Once again the orbital is skewed because of the difference in orbital energies. This time because of the lower energy of the fluorine $2p_z$ orbital (or, if you like, because of the higher electronegativity of fluorine relative to hydrogen) as opposed to the hydrogen 1s orbital, the electrons in the σ-molecular orbital have a higher probability of being found near the fluorine nucleus than the hydrogen nucleus.

The CO and NO Molecules

The similarity between carbon, nitrogen, and oxygen orbital energies is sufficient that the same molecular orbital scheme as for homonuclear diatomic molecules applies for

FIGURE 5.18. Formation of σ- and σ^*-molecular orbitals from hydrogen 1s and fluorine $2p_z$ atomic orbitals.

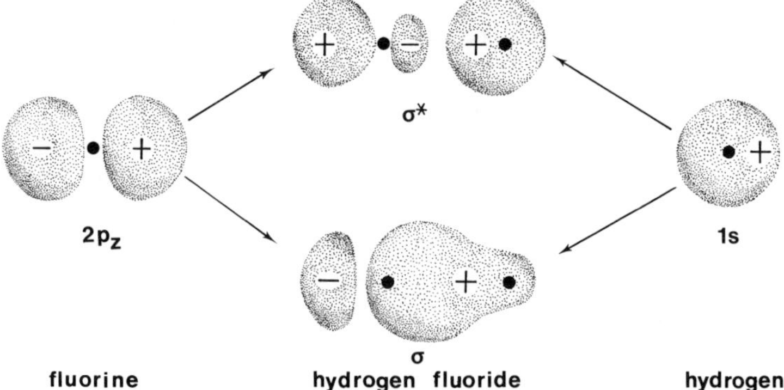

carbon monoxide and nitric oxide. We can write the configurations as:

$$\text{CO: KK}(2s_O)^2(2s_C)^2\pi^4\sigma^2$$
$$\text{NO: KK}(2s_O)^2(2s_N)^2\pi^4\sigma^2\pi^{*1}.$$

The bond orders are 3 and 2½, respectively. The extra unpaired electron in NO should cause NO to be paramagnetic, and this is indeed observed to be the case. The experimental bond energies of 11.06 eV molecule^{-1} and 7.00 eV molecule^{-1} for CO and NO are about what we would expect for a triple bond and a 2½ bond (cf. N_2 and O_2^+).

Experimentally Measurable Molecular Parameters

We have now seen how and why some atoms combine with others to form molecules with ionic bonds and how these lead to the formation of ionic crystals. We have also studied in some detail the formation of diatomic molecules by covalent bonding. Before proceeding with further theories of bonding for polyatomic molecules (and these theories will be even grosser approximations), we discuss now some experimentally determined molecular parameters that are useful in evaluating our theories.

There are a great many of these parameters that we could choose as tests, but we use only some or all of the following: (a) bond lengths, (b) interbond angles, (c) bond energies, (d) dipole moments, and (e) magnetism. All of these (except, of course, interbond angles) were used as criteria to test our theories for diatomic molecules in the previous sections. We want now to say a little bit more about what these particular quantities mean.

Bond Lengths

The bond length is the equilibrium distance between two nuclei connected by a bond (see p. 88). It is often remarkably constant for a given pair of nuclei from molecule to molecule. Table 5.4 contains some experimental data on

Table 5.4 Some experimental lengths of single bonds [in pm]

O—H		N—H		C—Cl		N—F	
H_2O	95.71	NH_3	101.36	CCl_4	176.9	NF_3	136.5
CH_3OH	95.6	NHF_2	102.6	$CHCl_3$	175.8	NHF_2	140.0
HCOOH	97.2	NH_2CN	100.1	CH_2Cl_2	177.2	FNNF	138.4
$HONO_2$	96.4	HNCO	98.7	CH_3Cl	178.1	FNO	152
		HN_3	97.5	CF_3Cl	174	FNO_2	183

single bonds illustrating the degree of variation commonly observed. These data were obtained by the experimental methods discussed in Chapter 7.

As we saw for diatomic molecules, double and triple bonds differ in length from single bonds. The same is true for polyatomic molecules. For example:

Bond	Length [pm]	Example
C—C	154	ethane
C═C	134	ethylene
C≡C	121	acetylene

Similarly, C—O bonds are in the range 139–143 pm, while C═O bond lengths are about 117–122 pm.

From an analysis of a large amount of experimental data on bond lengths, a table of covalent radii for various atoms can be derived, as we noted in Figure 4.15. The covalent radius is often taken as half the bond length when an atom is bonded to the same atom, or is obtained by various extrapolation procedures when molecules appropriate for this method do not exist. From these covalent radii tabulated in Table 5.5 bond lengths may be estimated. For

Table 5.5 Covalent radii for atoms [in pm]

	H	C	N	O	F
Single-bond radius	30	77	70	66	64
Double-bond radius		67	61	57	55
Triple-bond radius		60	55	51	
		Si	P	S	Cl
Single-bond radius		117	110	104	99
Double-bond radius		107	100	94	89
Triple-bond radius		100	93	87	
		Ge	As	Se	Br
Single-bond radius		122	121	117	114
Double-bond radius		112	111	107	104
		Sn	Sb	Te	I
Single-bond radius		140	141	137	133
Double-bond radius		130	131	127	123

example, the length of a C—Cl bond is calculated to be 77 + 99 = 176 pm, which is close to the observed length in many molecules containing this bond.

As mentioned in Chapter 4, the concept of the size of atoms is necessarily hazy, so we should not be too surprised if our use of covalent radii to predict bond lengths is sometimes inaccurate.

Interbond Angles

The angle, usually referred to as the *bond angle,* between two bonds to the same atom can also be determined experimentally (see again Chapter 7). Again for an atom bonded in a particular way the angles do not vary much from molecule to molecule. For example, HCH bond angles in the following molecules only vary over a few degrees.

Molecule	HCH Angle [deg]
CH_4	109.5
CH_3Cl	110.5
CH_2Cl_2	112
CH_3OH	109.3

The geometrical distribution of bonds in space obey a set of rules that is enunciated more fully later (pp. 135–139).

Bond Energies

The bond energy is simply the energy required to separate the component atoms of a bond. For diatomic molecules, as we have seen, the bond energy is called the dissociation energy, D_e. The bond energy is a useful measure of bond strength and can be correlated with bond order; in other words, the greater the bond order, the greater the bond energy.

For polyatomic molecules the definition sometimes may be ambiguous. For example, in methane, CH_4, it requires more energy to remove the first hydrogen than it does for successive ones. Thus the C—H bond energy in methane would be an average of four different quantities. For a given pair of atoms we expect the bond energy to be greater for a double bond than a single bond, and again greater for a triple bond than for either of these. Typical values for carbon-carbon bond energies are:

C—C	345 kJ mol^{-1}
C═C	610 kJ mol^{-1}
C≡C	835 kJ mol^{-1}

Our use of the bond-energy concept is limited to the idea that it correlates directly with bond order. Physical chemists often use bond energies as an empirical method of determining basic thermodynamic quantities.

Dipole Moments

If a molecule has a symmetrical charge distribution, that is, if there is no *net* separation of charge it is termed *nonpolar*. Familiar examples would be all homonuclear diatomic molecules and polyatomic molecules that have a highly symmetrical structure such as methane, carbon dioxide, ethane, benzene, and SF_6. However, as we have seen, most molecules have a nonsymmetrical charge distribution and are termed *polar*. Examples here would include all heteronuclear diatomic molecules, water, chloroform, ammonia, bromobenzene, and many others. The amount of polarity a molecule has is gauged by a quantity called the *dipole moment*. It is given the symbol μ and is defined as $\mu = QR$. These quantities are illustrated in Figure 5.19.

The units of the dipole moment are C m. Since for most molecules the value of the charge is of the order of the electronic charge and the separations are typically 100 pm, the value of μ will be of the order of:

$$(1.6 \times 10^{-19}\ \mathrm{C})(10^{-10}\ \mathrm{m}) = 1.6 \times 10^{-29}\ \mathrm{C\ m}.$$

This is rather cumbersome magnitude so an alternative, non-SI unit called the Debye (symbol D) has been almost universally used for molecular dipole moments. The conversion to SI units is

$$1\ \mathrm{D} = 3.3356 \times 10^{-30}\ \mathrm{C\ m}.$$

We can illustrate the use of dipole moments in gauging molecular properties by looking at the dipole moments of typical diatomic molecules given in Table 5.6.

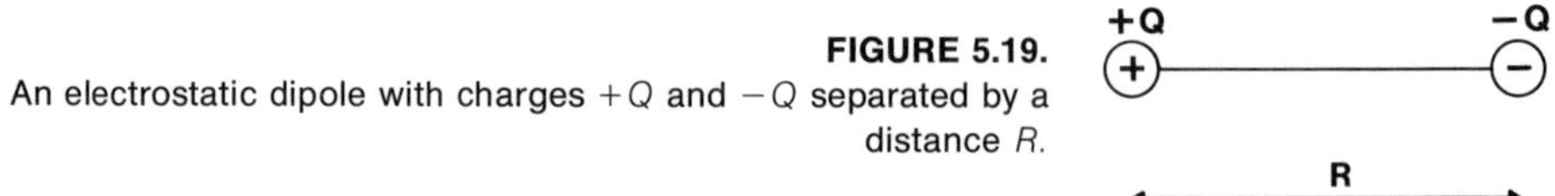

FIGURE 5.19. An electrostatic dipole with charges $+Q$ and $-Q$ separated by a distance R.

Table 5.6 Dipole moments of some diatomic molecules

Molecule	R [pm]	Calculated eR [C m × 10^{-30}]	Experimental μ [C m × 10^{-30}]	μ/eR [percent]
LiH	159.5	25.5	19.6	77
LiF	156.4	25.1	21.1	84
LiCl	202.1	32.4	23.8	74
LiBr	217.0	34.8	24.2	70
LiI	239.2	38.3	24.8	65
NaF	192.6	30.9	27.2	88
NaCl	236.1	37.8	30.0	79
KF	217.2	34.8	28.7	82
KCl	266.7	42.7	34.3	80
KBr	282.1	45.2	34.7	78
RbF	227.0	36.4	28.5	78
CsF	234.5	37.6	26.3	70
CsCl	290.6	46.6	34.8	75
HF	91.7	14.7	6.1	41
HCl	127.5	20.4	3.6	18
HBr	141.5	22.7	2.7	12
HI	160.9	25.8	1.5	6
SrO	192	61.5[a]	29.7	48
BaO	194	62.2[a]	26.5	43
AgCl	228.1	36.5	19.0	52

[a]For SrO and BaO each ion has a charge of $2e$; hence the calculated value is $2eR$.

In Table 5.6 the calculated dipole moment, eR, is obtained by assuming purely ionic bonding, that is, complete transfer of an electron from one nucleus to another. In all of the tabulated molecules this is an extreme approximation, as the experimentally observed dipole moment, μ, is always lower in value than that calculated. This means that there is a certain amount of electron sharing between the nuclei involved even though, for instance, for NaCl, we generally consider the bonding to be ionic. The final column gives the μ/eR as a percentage, which may be thought of as the *ionic character* discussed earlier (pp. 114–118).

Using molecular orbital theory, we say the ionic character reflected the skewedness of the molecular orbitals. An alternative description that perhaps is more useful for those molecules with a high degree of ionic character is to say that the cation *polarizes* the anion to some extent (see Figure 5.20).

The cation distorts and pulls the electron distribution of the anion toward it. This has the effect of sharing the electrons between the nuclei to some extent and thus

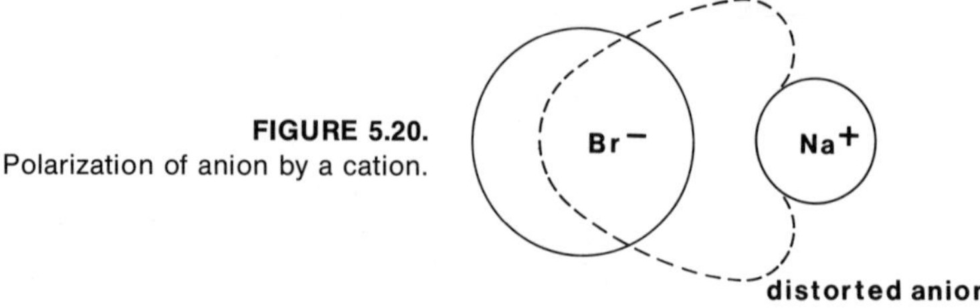

FIGURE 5.20. Polarization of anion by a cation.

reducing the dipole moment. When this occurs the measured dipole moment will always be smaller than that calculated for pure ionic bonding (touching spherical ions model). Yet another way of describing this phenomenon by resonance is given later (p. 134).

The data in Table 5.6 show a number of other features. Firstly, the values of μ/eR vary from 88 percent for NaF to 6 percent for HI, indicating the extremes of the range for nearly pure ionic bonding to nearly pure covalent bonding (of course, for homonuclear diatomic molecules there is no dipole moment and the bonding is essentially covalent). Other molecules in the table are somewhere in between. There is also a broad generalization that comes from considering more data than those included in Table 5.6:

> The extent of the polarization, that is, the degree of covalent character of the bonds, depends on the nature of the cation and the anion. The larger the anion, the less strongly the valence electrons are bound and the more easily the anion is polarized. Also the greater the charge on the cation, the stronger is its polarizing effect on a given anion. Furthermore, the smaller the cation for a given charge, the closer it can approach the anion and so the greater is its polarizing ability.

Thus when a large anion is combined with a highly charged or small cation, we expect fo find that the bonding has considerable covalent character. In compounds such as $FeCl_3$ or HCl the bonds turn out to be predominantly covalent. In Table 5.6 the data on the hydrogen halides lend support to this comment. The data on SrO and BaO can be taken to indicate the effect of having highly charged atoms.

Magnetism

We do not use the magnetic properties of molecules to any great extent here to test our theories of valency except to say whether a molecule is paramagnetic or diamagnetic. You will recall that paramagnetism arises because of unpaired electrons, whereas diamagnetism indicates that all of the electrons have paired spins. Thus we want our valence theories only to answer "yes" or "no" to the question of whether there are any unpaired electrons in the molecule. In the majority of cases this answer will be "no" because most common chemical species are diamagnetic.

Electron Sharing and the Completed Valence-shell Principle

In the case of more complicated molecules involving three or more nuclei, we can also trace the formation of molecular orbitals as the constituent atoms come together. However, while it is still fruitful to represent these molecular orbitals by linear combinations of atomic orbitals, the picture gets rather more complicated, and we do not pursue the story for the present. Instead we begin our discussion of bonding in polyatomic molecules by returning to the simpler picture of electron sharing introduced by G. N. Lewis before quantum mechanics was properly developed. Lewis pointed out that, for a very great number of molecules for which the concept of ionic bonding failed to account for the observed stability, an empirical principle based on the sharing of valence electrons to complete the valence shell would uniformly account for the observed structures and their stability. He specifically proposed that, when two or more atoms come together and arrange their valence electrons so that some of them are shared between two nuclei, the system is a stable molecule *if the sharing occurs in such a way that all nuclei achieve a completed valence shell of electrons.*

For a hydrogen atom this principle means sharing *two* valence electrons with some other atom. For the elements Li through Mg it is *eight* (sometimes referred to as an octet), and for elements beyond Mg it is often eight but can be as many as 18. For these elements the valence shell may include the d-orbitals. Thus the valence shell may or may not be expanded from the one s- and three p-orbitals (eight electrons) to include the d-orbitals (10 more electrons giving a total of 18 in the valence shell).

There is a close relationship between the shared pair of electrons that Lewis used to form a covalent bond and a pair of electrons occupying a bonding molecular orbital such as the $1s\sigma$ orbital in H_2. To indicate how the valence

FIGURE 5.21. Representation of the F_2 molecule by the dot (a) and dash (b) notations.

electrons are distributed around a molecule, two notations are widely used. The first uses dots to indicate valence electrons as in Figure 5.21(a), and the second uses a dash to represent a pair of electrons[11] as in Figure 5.21(b). Notice that some of the valence electrons are left as pairs on a single nucleus not shared between two nuclei. Such unshared pairs are referred to as *lone pairs*.

It is possible for more than one pair of electrons to be shared. Thus in order to complete valence shells, ethylene and nitrogen are written as in Figure 5.22. Here two pairs of electrons are shared between the two carbon atoms in ethylene and three pairs are shared between the two nitrogen atoms in N_2, giving rise to double- and triple-bond representations.

By devising sharing processes to complete valence shells we may write the electronic structure of many molecules as in Figure 5.23.

When covalent bonding by completing valence shells is used to write electronic structural formula, each atom participates in a definite number of bonds. This number is referred to as the atom's *valency*. Table 5.7 lists some common valencies.

When we consider the electronic structure of molecular ions and a few other molecules it is necessary to write

[11]Note that the term *pair* is now widely used to mean two electrons occupying the same orbital and thus having opposite spins.

FIGURE 5.22. Lewis structures showing the double bond in ethylene and triple bond in nitrogen.

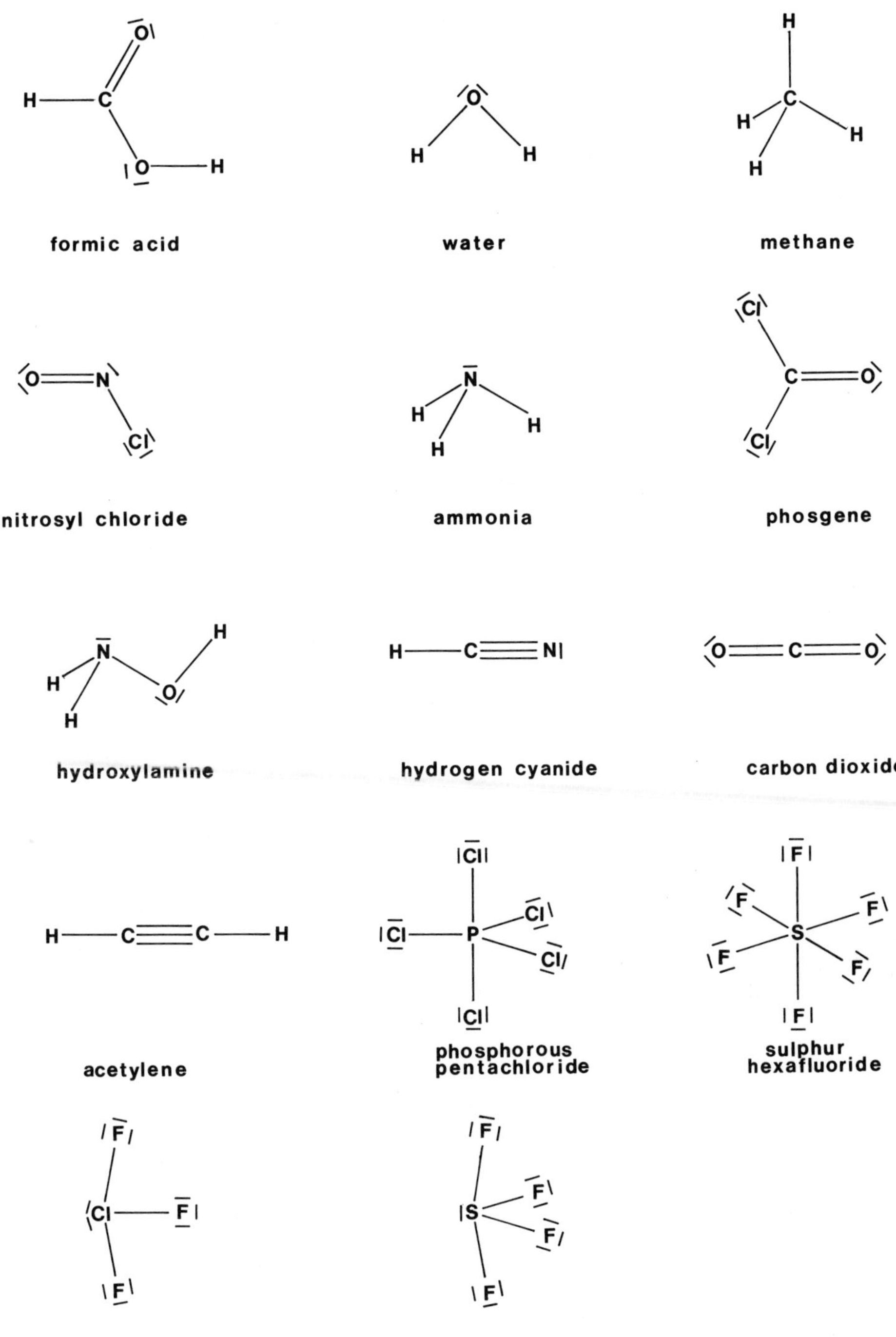

FIGURE 5.23. Structural formulas for some covalent molecules.

Table 5.7 Covalent valencies

Atom	Number of Valence Electrons	Valency
H	1	1
C	4	4
N	5	3
O	6	2
F	7	1
P	5	3 or 5[a]
S	6	2, 4 or 6[a]
Cl	7	1, 3 or 5[a]

[a]Not all of these valencies are illustrated in Fig. 5.23.

structures with so-called *formal charges* on some of the atoms involved. For example:

$\ominus$ |C≡N|

cyanide ion

H_3C—C≡N$^{\oplus}$—$\bar{O}|^{\ominus}$

acetonitrile N-oxide

A general rule for deriving the formal charge (FC) on an atom is:

> The formal charge is equal to the number of valence electrons of the neutral atom minus the number of lone-pair electrons minus half the number of electrons involved in covalent bonds.

Thus for the nitrogen atom in CH_3CNO:

$$FC = 5 - \frac{1}{2}(8) = +1$$

and for oxygen:

$$FC = 6 - 6 - \frac{1}{2}(2) = -1.$$

Other examples are:

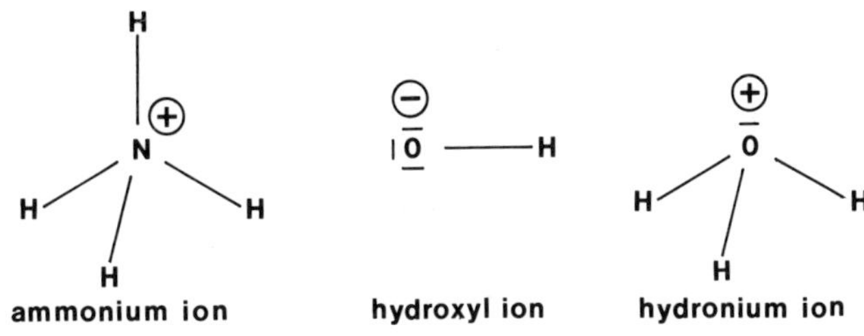

We next see that for some molecules for which it is necessary to invoke the concept of resonance to write the electronic structural formula, structures with formal charges are often involved.

Resonance

In the last section we derived formulas to represent the electronic structure of various molecules and ions using the completed valence-shell principle to satisfy the valency requirements of each atom in the molecule. Unfortunately, life is not quite so simple because for some molecules and molecular ions it is possible to satisfy the valency requirements in more than one way. Consider, for example, the nitrous oxide molecule. We may write three seemingly reasonable structural formulas (Figure 5.24). These three structural formulas obviously correspond to different distributions of the valence electrons. In Figure 5.24 the numbers below the structures represent the number of lone-pair and bonding electrons on the atoms and in the bonds.

FIGURE 5.24.
Possible structural formulas (canonical forms) for nitrous oxide. The numbers represent lone pair and bond electrons.

What, then, is the actual structure of nitrous oxide? It is a weighted average or *resonance hybrid* of the possible structures called *canonical forms* or *resonance structures*. So too the bond orders are the same sort of weighted average of the bond orders in the resonance structures.

It is very important for the student to understand that there is no physical reality to the concept of resonance. It is merely a crutch we introduce to conceal the inadequacies of our very crude valency theory of representing bonds by dashes. Do not be misled into thinking a particular resonance hybrid can be physically or chemically isolated from others. There is only one kind of nitrous oxide molecule, but we have to use more than one picture, then average the pictures, to describe its bonding properties in a reasonable fashion.

When expressing resonance structures in writing some will often be preposterous. As we gain experience in the principles of valency we can deduce just which resonance structures should be included. Some simple guidelines to follow are:

1. Structures should obey the completed valence-shell principle.
2. Those structures where a given atom has a multiple formal charge can be omitted (i.e., their weight in the weighted average is very small). Structure (c) for nitrous oxide in Figure 5.24 could be omitted on this basis.
3. Those structures with like charges on adjacent atoms can be omitted. Simple electrostatics is the guideline here; again omit structure (c) in Figure 5.24.
4. The best structures—those that contribute the most to the actual structure—are those in which the number of formal charges is kept to a minimum.

For sake of argument, let us assume the actual weighted average of nitrous oxide to be represented by

$$\text{Actual structure} = 0.45\text{a} + 0.54\text{b} + 0.01\text{c}$$

where a, b, and c refer to the respective structures in Figure 5.24. Bond-length data from experiment show for nitrous oxide

NN	bond	112pm
NO	bond	119 pm.

The possible calculated bond lengths for these bonds from Table 5.5 are

N≡N 110 pm
N=N 122 pm
N=O 118 pm
N—O 136 pm.

Thus we see that the experimental values are in between a double and triple bond for the NN bond and in between a single and double bond for the NO bond.

There are many molecules that can be described reasonably accurately using the concept of resonance. One in which the structure is an exact average of the canonical forms because of symmetry requirements is the carbonate ion CO_3^{2-} illustrated in Figure 5.25. Here we have used the double headed arrow ↔ to indicate that the actual structure is an average of the three structures (not to be confused with the equilibrium symbol ⇄). Notice that all of the bonds for the carbonate ion should be equivalent and the bond order should be exactly 1⅓ because the resonance structures contribute equally because of symmetry. The experimental CO bond length in CO_3^{2-} is 129 pm, which supports this description because it is between the C—O and C=O distances calculated from covalent radii of 143 and 124. pm respectively.

Another example is nitric acid, where the contributing resonance forms can be written as in Figure 5.26. Notice that in order to write structural formulas we must know beforehand which atoms are bonded to each other. Furthermore, we do not include structures such as:

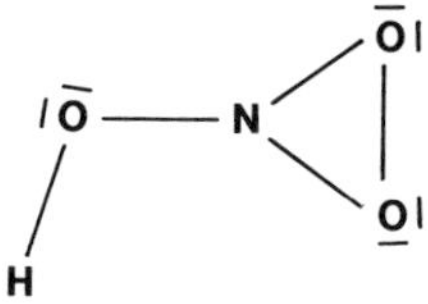

for nitric acid since the experimentally observed OO bond distance is 220 pm, and this is much greater than twice the

FIGURE 5.25. Resonance structures of the carbonate ion.

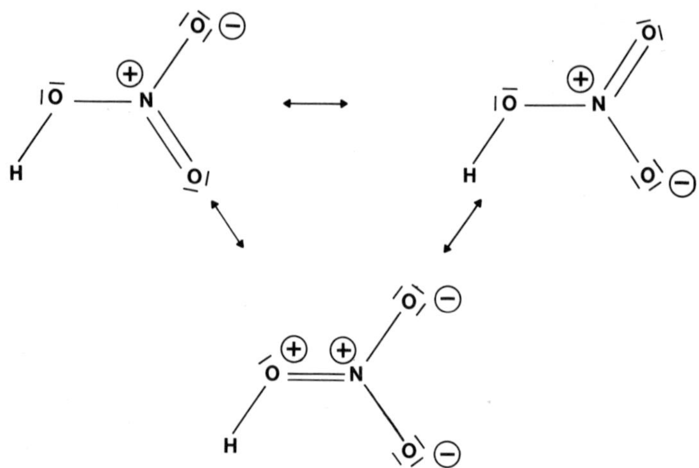

FIGURE 5.26. Resonance forms for nitric acid.

single-bond covalent radius of oxygen (2 × 66 = 132 pm). The NO bond distance is 124 pm, between the double and single bond distance, as we would expect from Figure 5.26. Notice that the third structure in Figure 5.26 is not preferred because of the multiplicity of formal charges and adjacent like formal charges. On these grounds it would probably contribute little to the resonance hybrid. Finally,

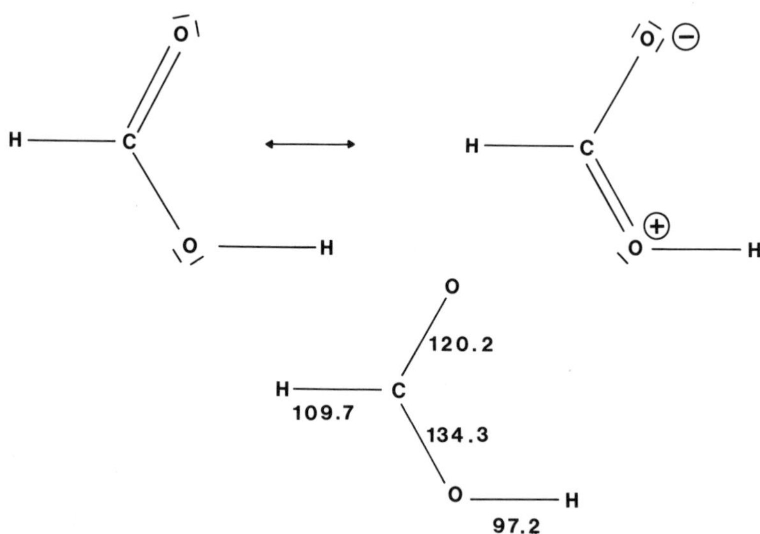

FIGURE 5.27. Resonance structures and bond lengths for formic acid.

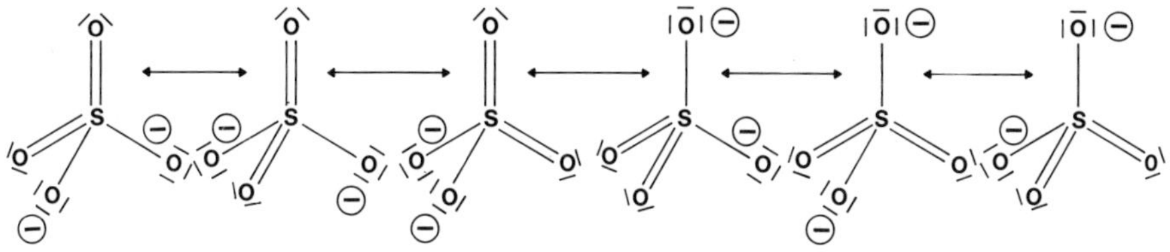

FIGURE 5.28. Resonance structures for the sulfate ion.

we can conclude from the resonance forms showing formal charges that nitric acid should be a polar molecule and possess an appreciable dipole moment. In fact, the measured dipole moment is 2.17 D. For the carbonate ion, on the other hand, although each CO bond would have dipolar character, they would cancel overall because the dipole moment is a *vector* quantity and three equal vectors directed at angles of 120° to each other cancel.

Yet another example of resonance is illustrated by the formic- acid molecule, HCOOH. The two major resonance structures and experimental geometry are given in Figure 5.27. Here one of the CO bond lengths is between the values for C—O (147 pm) and C═O (124 pm). The other is even shorter than a standard C═O bond length. This merely emphasizes that we must beware of pushing our simple theory too hard. There will be exceptions, as the formic acid bond lengths indicate. Probably in this case the structure placing the positive formal charge on the oxygen is of little importance.

Figure 5.28 shows six reasonable resonance forms for the sulfate anion SO_4^{2-}.

We exclude possible structures where the valency of the sulfur atom exceeds six, such as:

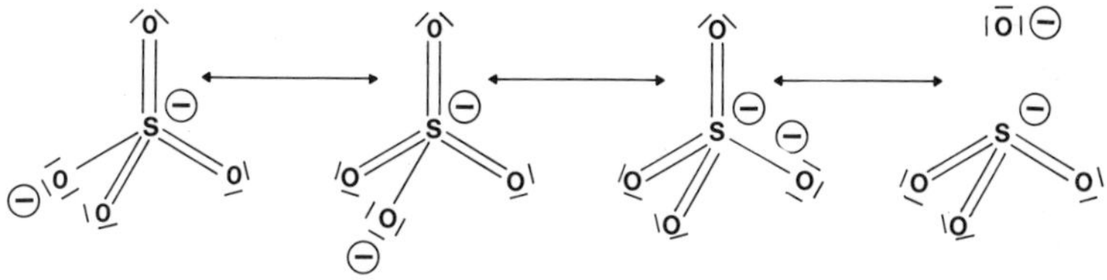

Notice that we have also omitted the structures that have multiple charges on sulfur and more charges than the minimum number possible, e.g.:

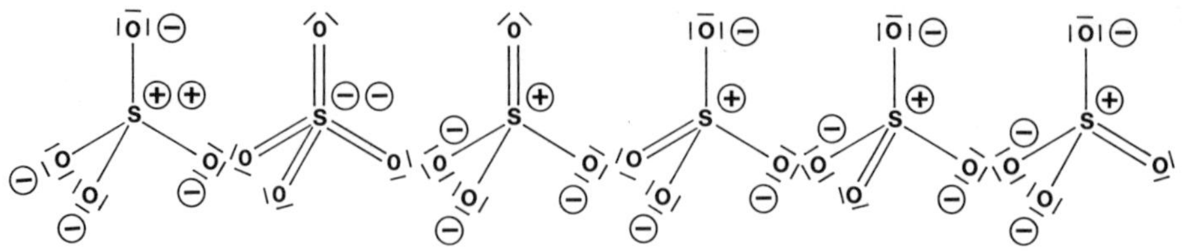

Furthermore, the resonance hybrid of the six structures in Figure 5.28 would have an SO bond order of 1.5. The experimental bond length is 150 pm (this number is an average of many values taken from X-ray data on ionic sulfates), but this is nearly identical with the standard double bond length (see Table 5.5), again illustrating the limitations of our simple bonding theories.

One final example to illustrate the concept of resonance in polyatomic molecules is the sulfur dioxide molecule, the resonance structures of which are shown in Figure 5.29. For the reasons discussed structure (a) can be discarded. Equivalent structures (b) and (c) indicate a polar molecule, and structure (d) says that the bond order should be greater than 1.5. The dipole moment of 1.64 D and bond length of 143 pm support this picture of the SO_2 molecule.

Before leaving this section on resonance we want to show how the concept can also be used to bridge the gap between ionic and covalent bonding in diatomic molecules. Consider, for example, the case of hydrogen chloride. We can write two structures

$H^+ \;|\overline{\underline{Cl}}|^-$	$H - \overline{\underline{Cl}}|$.
Ionic bond (complete transfer of one valence electron from H to Cl)	Covalent bond (equal sharing of two valence electrons between H and Cl)

The principle of resonance indicates that the actual electronic structure is a weighted average of these two extremes; that is, the bond in HCl is neither purely ionic nor purely covalent. The experimental bond length in the HCl molecule is 127.5 pm, which is very close to the sum of the covalent radii of 129 pm. The ionic radius of Cl^- is

FIGURE 5.29. Resonance structures of SO_2.

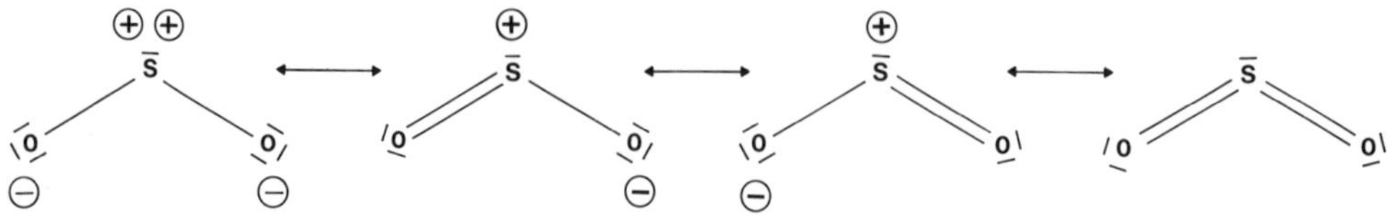

181 pm, so that the bond-length data imply that HCl is much closer to the covalent than the ionic structure. The data on dipole moments (Table 5.6) also predict a predominantly covalent molecule.

Geometric Distribution of Bonds in Space

Thus far we have cunningly drawn diagrams of molecular electronic structure formulas roughly in the form of their geometric distribution in space. We have not discussed the reasons why molecules take the particular shapes that they do. In this section we reinforce some elementary ideas about predicting shapes of molecules that should be already familiar. In the next section (on valence-bond theory) we discuss a more sophisticated (but not necessarily more satisfactory) model of explaining bond angles in molecules through a concept known as *hybridization*.

The simplest theory to explain the angles between the bonds about an atom was proposed in 1939 by N. V. Sidgwick and H. M. Powell. Known as the Sidgwick–Powell rule, it is:

> The valence electrons in bonds and lone pairs repel one another and so the bonds and lone pairs tend to keep as far apart as possible.

Notice one important fact however, we need to know beforehand which atoms are bonded to which, that is, the nuclear framework, before we can apply this principle.

To satisfy the Sidgwick–Powell rule, two electron pairs will remain 180° apart, three will remain 120° apart in a plane, and four will adopt a regular tetrahedron arrangement [bond angles 109.5°, i.e., arc cos $(-1/3)$]. When there are five pairs we get a trigonal bipyramid arrangement and finally six pairs give an octahedral geometry. These shapes with the different possibilities for lone pairs or bonding pairs are shown in Figure 5.30.

We now list some simple examples that illustrate the rules with bond pairs:

1. $HgCl_2$ has two bond pairs and is linear.

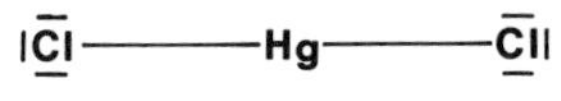

2. BCl_3 has three bond pairs and is plane–triangular.

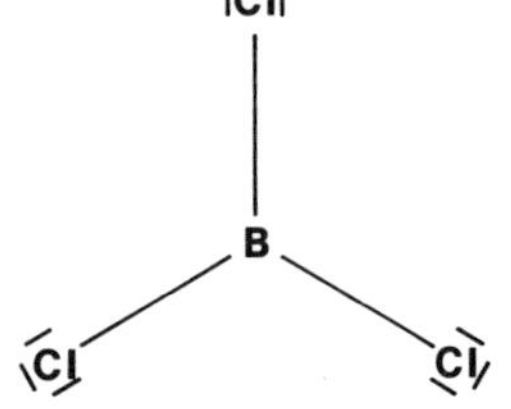

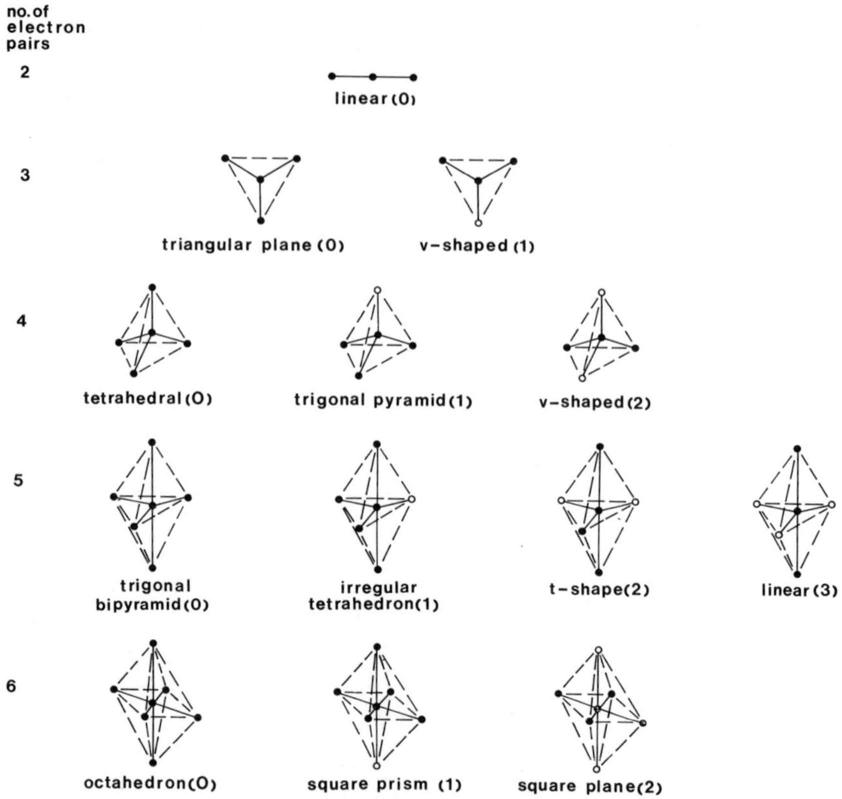

no. of electron pairs
2
linear (0)
3
triangular plane (0)
v-shaped (1)
4
tetrahedral (0)
trigonal pyramid (1)
v-shaped (2)
5
trigonal bipyramid (0)
irregular tetrahedron (1)
t-shape (2)
linear (3)
6
octahedron (0)
square prism (1)
square plane (2)

FIGURE 5.30. Shapes of molecules (nontransitional elements) (numbers of lone pairs given in parentheses and indicated by open circles).

3. CH_4 has four bond pairs and is tetrahedral.

H
C
H
H
H

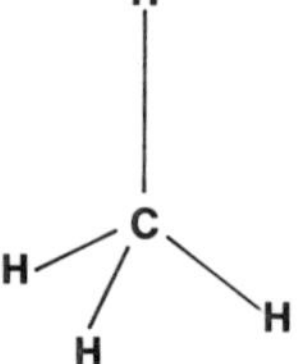

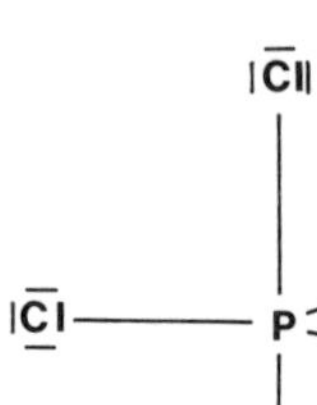

4. PCl_5 has five bond pairs and is trigonal bipyramidal (note for this molecule the two axial bonds are not symmetrically equivalent to the three equatorial bonds).

Cl
Cl
P
Cl
Cl
Cl

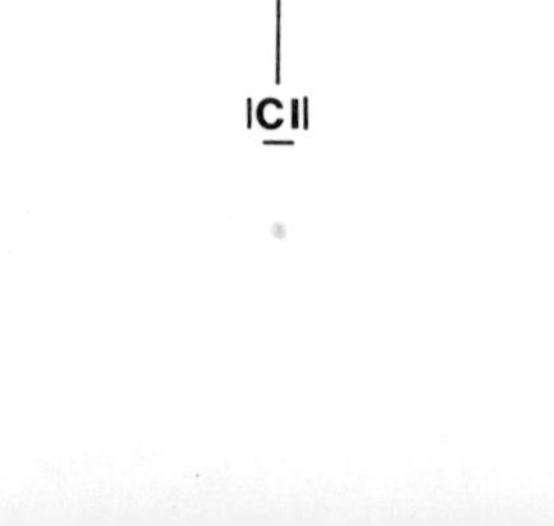

5. SF_6 has six bond pairs and is octahedral.

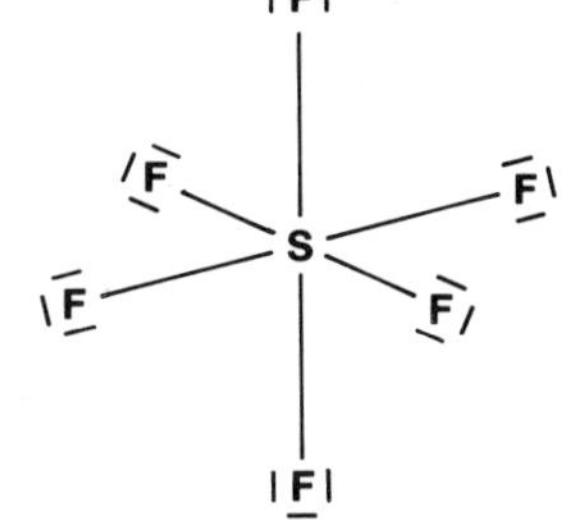

When one or more of the pairs of valence electrons is a lone pair, this commands a greater solid angle than a bonding pair. An important example is the ammonia molecule. The lone pair on nitrogen takes more than one quarter of the complete solid angle around the nitrogen, each N—H bonding pair occupying less than one quarter. Thus the HNH angle is less than the regular tetrahedral value by about 3°. Similarly for water, the two lone pairs will occupy more than half the total solid angle around oxygen. Thus we predict that the HOH angle will deviate even further from the tetrahedral value. It does, and experiment shows the angle to be 104.5°. Table 5.8 lists some examples and structures from which a similar deduction may be made are shown in Figure 5.31.

When there are multiple bonds present, such as double bonds or resonance hybrid structures where the bond order may be nonintegral but greater than 1, the same principles apply as in the case of single bonds. Thus in the absence of lone pairs, two double bonds will result in a linear molecule, while three double bonds in a molecule cause it to be plane–triangular in shape (see Figure 5.32). Similarly, NO_3^- and CO_3^{2-} will be plane–triangular.

If now there is a combination of single and multiple bonds, simple repulsion arguments predict that the multiple bonds will require more solid angle than the single

Table 5.8 Some experimental bond angles

Molecule	Bond angle [deg]	Molecule	Bond angle [deg]
NH_3	106.6	H_2O	104.5
NF_3	102.9	F_2O	103.2
PCl_3	100.1	H_2S	92.2
$AsCl_3$	98.4	H_2Se	91.0
ClO_3^-	106.7	ClO_2^-	110.5

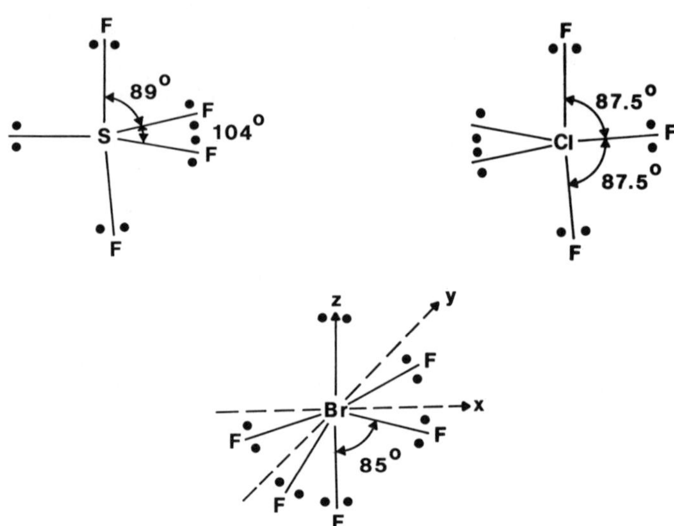

FIGURE 5.31. Geometries of SF_4, ClF_3, and BrF_5.

bonds. Consider, for example, the nitric-acid molecule in Figure 5.33. The bond lengths in this planar molecule suggest that the N—OH bond is effectively a single bond, whereas the two N—O bonds have a bond order of about 1.5 (see Figure 5.26). The bond angles show clearly that

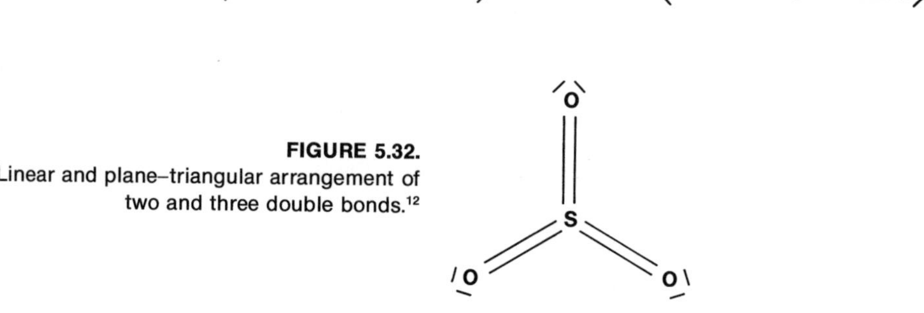

FIGURE 5.32. Linear and plane–triangular arrangement of two and three double bonds.[12]

[12]To keep ourselves honest with structures that are as complete as possible, we should include the resonance forms:

FIGURE 5.33. The bond angles in nitric acid.

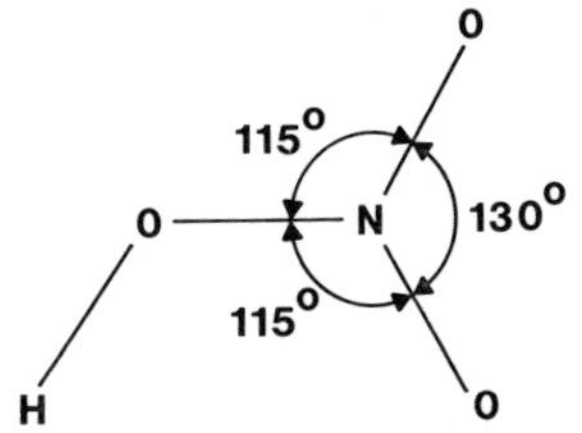

these bonds repel each other more than they repel the N—OH bond. We can use similar arguments to explain the geometries of other molecules with multiple bonds such as in Figure 5.34.

Finally, when there are multiple bonds and lone pairs, the experimental geometries indicate that the lone pairs are still the most space-hungry valence electrons, even more so than double and triple bonds. Some examples are given in Figure 5.35.

FIGURE 5.34. Geometries of some molecules with single and multiple bonds.

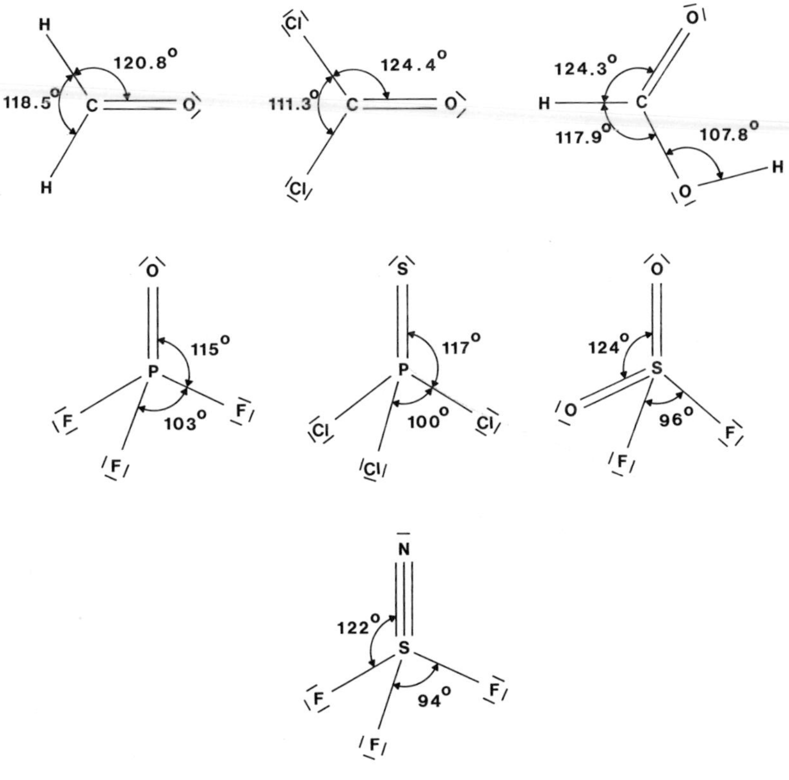

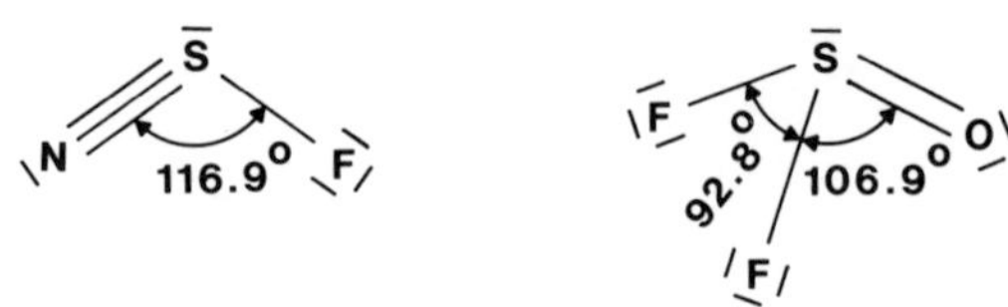

FIGURE 5.35. Geometries of some molecules with lone pairs and multiple bonds.

Valence-bond Theory and Valence States

The Sidgwick–Powell rule is an excellent tool for predicting molecular geometries, but it does not make use of the details of atomic structure that are portrayed by the atomic orbitals. One of our main aims in this book is to relate properties of molecules to the underlying electronic structure and so we must consider how the geometric features are to be related to atomic orbitals. The solution is to elaborate the concept of electron sharing, as described earlier (pp. 125–129), by considering more fully the atomic orbitals that are involved. This approach to the description of chemical bonds was developed largely by Linus Pauling and is termed the *valence bond theory*.

In this theory we imagine that a covalent bond is formed between two atoms by starting with an unpaired electron in an atomic orbital on each atom. The atoms are brought together so that the orbitals overlap and the spins of the two electrons become paired. The electrons might now be regarded as occupying *a localized molecular orbital*, located around the two nuclei that are bonded.

For example, we can picture the formation of H_2 by starting with two hydrogen atoms, each having its unpaired electron in its 1s orbital. The atoms are brought together, the orbitals overlap strongly (just as described on pp. 101–104 when we were considering molecular orbitals) and the electrons form a pair and are then "shared" between the two nuclei.

The valence bond approach places emphasis on pairing of spins. For example, in the case of the water molecule we imagine the two oxygen valence electrons that are forming bonds as occupying one oxygen atomic orbital each so that they can pair off with a hydrogen electron in each case (see Figure 5.36).

However, in many instances the atom, if it is in the most stable state, does not have the appropriate number of unpaired electrons to form bonds by spin pairing. For example, in the case of beryllium the most stable state (the ground state) has the electron configuration $(1s)^2(2s)^2$ with all spins paired. Nevertheless, it forms compounds such as

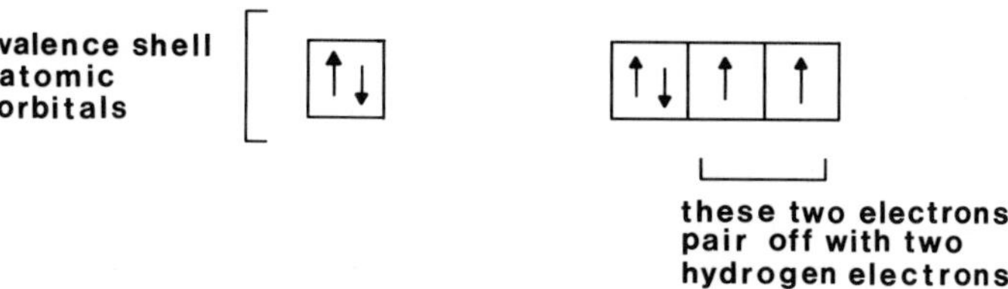

FIGURE 5.36. Oxygen valence electrons.

BeH_2 with two covalent bonds emanating from the beryllium. From the valence-bond viewpoint, therefore, we need to consider a different electronic configuration of the atom in which two valence orbitals are singly occupied, the one of lowest energy being $(1s)^2(2s)^1(2p)^1$. Such a state is referred to as the *valence state* of the atom (see Table 5.9).

Hybridization In compounds such as BeH_2 the two bonds formed by the beryllium are identical in all of their properties. If we were to associate one of the bonds with a beryllium electron in the 2s orbital and the other with the electron in the 2p orbital, we might surmise that the bonds were not in fact identical. Fortunately, quantum-mechanical principles show us that it is a matter of taste how one dissects the electron distribution into contributions from individual electrons. One way of distributing the electrons so as to produce the same total distribution is to place one electron into each of the orbitals shown in Figure 5.37. This new description proves to have two advantages over the $(2s)^1$ $(2p)^1$ description. Firstly, the two orbitals have the same shape as each other and, secondly, each orbital is largely located in the region of one of the BeH bonds.

To obtain these new orbitals we mathematically com-

Table 5.9 Atomic ground states and valence states

Atom	Electron configuration of ground state	Electron configuration of valence state	Abbreviation (unpaired electrons)
Li	$(1s)^2$ (2s)	Same as ground state	s
Be	$(1s)^2$ $(2s)^2$	$(1s)^2$ (2s) (2p)	sp
B	$(1s)^2$ $(2s)^2$ (2p)	$(1s)^2$ (2s) $(2p)^2$	sp^2
C	$(1s)^2$ $(2s)^2$ $(2p)^2$	$(1s)^2$ (2s) $(2p)^3$	sp^3
N	$(1s)^2$ $(2s)^2$ $(2p)^3$	Same as ground state	p^3
O	$(1s)^2$ $(2s)^2$ $(2p)^4$	Same as ground state	p^2
F	$(1s)^2$ $(2s)^2$ $(2p)^5$	Same as ground state	p

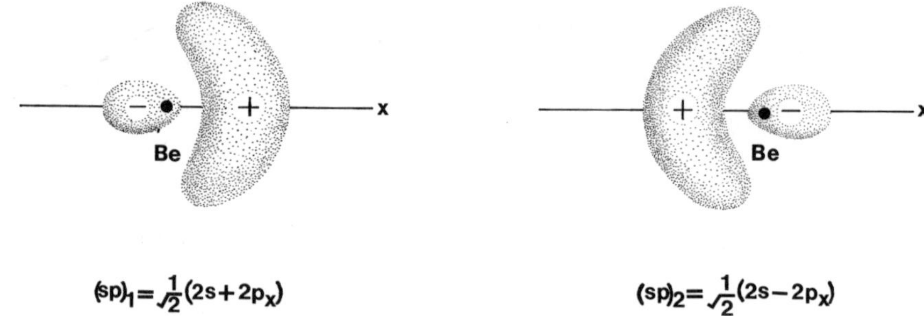

$(sp)_1 = \frac{1}{\sqrt{2}}(2s + 2p_x)$ $(sp)_2 = \frac{1}{\sqrt{2}}(2s - 2p_x)$

FIGURE 5.37. sp-Hybrid orbitals. Note that the orbitals are both centered on the same atom.

bine the 2s and $2p_x$ orbitals by taking linear combinations of them, namely:

$$\left.\begin{aligned}(sp)_1 &= \frac{1}{\sqrt{2}}(2s + 2p_x)\\ (sp)_2 &= \frac{1}{\sqrt{2}}(2s - 2p_x).\end{aligned}\right\} \tag{5.12}$$

These new functions are called *sp-hybrid orbitals* and the combining process is known as *hybridization of atomic orbitals*. Since the resulting electron density for putting one electron in each of the 2s and 2p atomic orbitals is the same as for putting one in each of the sp hybrids,[13] we are at liberty to choose the hybrids to represent the state of the Be atom prior to bond formation. The contour diagrams of the sp-hybrid orbitals are shown in Figure 5.37. It is obvious from the diagram that the two sp hybrid orbitals are directed in directions at 180° to one another; thus we may describe the bonding in BeH_2, knowing that the molecule will have a linear geometry. From the molecular-orbital viewpoint we now form localized σ-type molecular orbitals by taking linear combinations of each sp hybrid with a hydrogen 1s orbital. We get, as expected, four molecular orbitals, two from each combination of an sp hybrid with the hydrogen 1s orbital.

[13]We can show that the electron density is the same since

$$\begin{aligned}&\int_{-\infty}^{\infty}\left\{\left|\frac{1}{\sqrt{2}}(2s + 2p_x)\right|^2 + \left|\frac{1}{\sqrt{2}}(2s - 2p_x)\right|^2\right\} dv\\ &= \int_{-\infty}^{\infty}\{\tfrac{1}{2}|2s|^2 + \tfrac{1}{2}|2p_x|^2 + \tfrac{1}{2}|2s|^2 + \tfrac{1}{2}|2p_x|^2\}dv\\ &= \int_{-\infty}^{\infty}\{|2s|^2 + |2p_x|^2\}dv\end{aligned}$$

The cross terms are zero because $|2s|\,|2p_x|$ is an odd function; hence its integral from $-\infty$ to ∞ is zero.

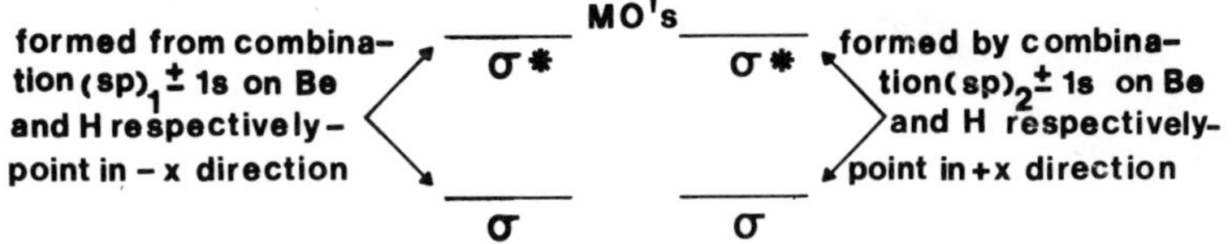

The four valence electrons of BeH_2 are then fed into these molecular orbitals and will occupy only the σ-molecular orbitals. This gives a description of the two covalent bonds directed properly in space.

When forming hybrid orbitals we must start with orbitals of similar energy, otherwise the corresponding valence state is of much higher energy than the ground state and so is unlikely to be involved in chemical bonding. For beryllium the $(1s)^2$ (2s) (2p) configuration is about 3.36 eV higher in energy than the $(1s)^2$ $(2s)^2$ state. As illustrated in Figure 5.38, the energy increase involved in going to the valence state is more than compensated by the lowering in energy that results from formation of two chemical bonds.

There are other combinations of atomic orbitals from which we can form other sets of hybrid orbitals with convenient directional properties. Two of these, which are constantly used in organic chemistry, are sp^2 and sp^3 hybrid orbitals.

sp^2-Hybrid Orbitals

These orbitals are constructed from linear combinations of an s orbital and *two* of the p-orbitals. As was implicit in our formation of sp hybrids, the s and two p-orbitals must have similar energies; hence they will always have the same *n* quantum number, namely, 2s and 2p, *or* 3s and 3p,

FIGURE 5.38. Energy relationships for beryllium atom in valence state and after forming two covalent bonds.

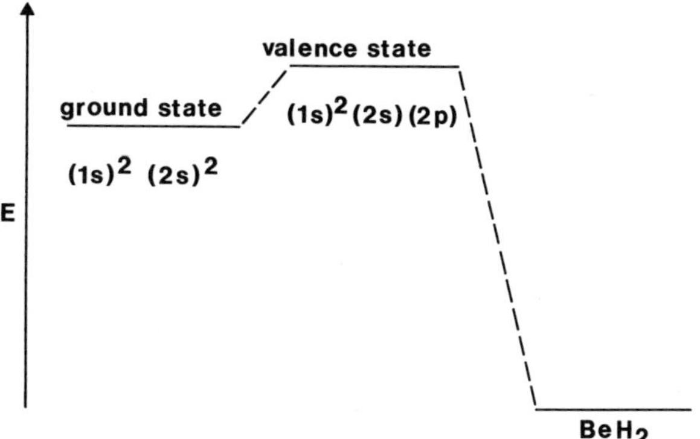

but not 2s and 3p. The mathematical forms of the sp^2 hybrids are:

$$\left.\begin{aligned} (sp^2)_1 &= \frac{1}{\sqrt{3}}\,s + \frac{2}{\sqrt{3}}\,p_x \\ (sp^2)_2 &= \frac{1}{\sqrt{3}}\,s - \frac{1}{\sqrt{6}}\,p_x + \frac{1}{\sqrt{2}}\,p_y \\ (sp^2)_3 &= \frac{1}{\sqrt{3}}\,s - \frac{1}{\sqrt{6}}\,p_x - \frac{1}{\sqrt{2}}\,p_y. \end{aligned}\right\} \qquad (5.13)$$

They are all equivalent and directed at 120° angles to each other in the x,y plane (since we have arbitrarily used the p_x and p_y orbitals). The sp^2 hybrids are shown diagramatically in Figure 5.39.

An example of the use of these hybrid orbitals would be in the description of BF_3. Each of the sp^2 hybrids on the boron atom forms a bonding σ-type and antibonding σ^*-type molecular orbital with an appropriate 2p orbital from each fluorine atom:

$$3\{sp^2(\text{boron}) \pm 2p(\text{fluorine})\} \begin{array}{l} \nearrow 3\sigma^* \\ \searrow 3\sigma \end{array} \quad \text{molecular orbitals}$$

pointed 120°
to each other.

The six valence electrons (three from boron and one from each fluorine) are then fed into the three bonding molecular orbitals, two in each to give the description of the bonding in BF_3.

In many molecules the bond angles may be a little less or a little more than 120°. Examples of formaldehyde, CH_2O, and phosgene CCl_2O were shown in Figure 5.34. In these cases the mathematical form of the sp^2 hybrids (Equation 5.13) will be slightly different, but for descrip-

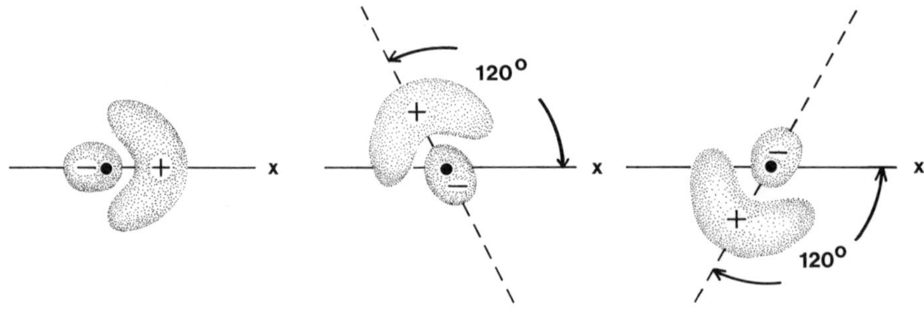

FIGURE 5.39. sp^2-Hybrid orbitals.

tive purposes this is not worth pursuing. In other words, we can use the sp^2 hybrid orbitals in these cases knowing that they are not quite exact.

sp^3 Hybrid Orbitals

The other very useful form of hybridization is sp^3. Here we mix one s and three p orbitals in the following way to form four sp^3 hybrid orbitals:

$$\begin{aligned}(sp^3)_1 &= \tfrac{1}{2}(s + p_x + p_y + p_z)\\ (sp^3)_2 &= \tfrac{1}{2}(s - p_x - p_y + p_z)\\ (sp^3)_3 &= \tfrac{1}{2}(s + p_x - p_y - p_z)\\ (sp^3)_4 &= \tfrac{1}{2}(s - p_x + p_y - p_z).\end{aligned} \tag{5.14}$$

These sp^3 hybrids have a shape similar to those of the sp and sp^2 hybrids in Figures 5.37 and 5.39, except they are directed toward the corners of a regular tetrahedron. Again we emphasize that the total charge distribution resulting when there is one electron in each of the four tetrahedral sp^3 hybrids is exactly the same as that when there is one electron in the s orbital and one in each of the p_x, p_y, and p_z orbitals. Furthermore, this total charge distribution is spherically symmetric. This last point is not always easy to visualize as there is a tendency when drawing diagrams to overemphasize the directional characteristics of hybrid orbitals because of the artistic difficulties of representing orbitals that heavily interpenetrate one another.

Other Hybrid Orbitals

Further hybrid orbitals are possible using d orbitals as well as the s and p orbitals. Table 5.10 lists some of the possible combinations that give sets of equivalent hybrids. One could use d^2sp^3 hybrids to explain the bonding in SF_6. A

Table 5.10 Equivalent hybrid orbitals

Combination	Symbol	Hybrids	Angle between hybrids
s, p_x	sp	Two diagonal	180°
s, p_x, p_y	sp^2	Three plane triangular	120°
s, p_x, p_y, p_z	sp^3	Four tetrahedral	109.5°
p_z, d_{z^2}	pd	Two diagonal	180°
s, p_x, p_y, $d_{x^2-y^2}$	dsp^2	Four in a plane	90°
s, p_x, p_y, p_z, d_{z^2}, $d_{x^2-y^2}$	d^2sp^3	Six octahedral	90°

combination of pd diagonal hybrids and sp^2 trigonal hybrids represents a set of five hybrid orbitals (not all equivalent) directed toward the corners of a trigonal bipyramid. This combination could be used to describe the electronic structure of a molecule such as PCl_5 as shown in Figure 5.40. The axial bonds involve pd hybrids (formed from phosphorus $3d_{z^2}$ and $3p_z$ atomic orbitals), whereas the equatorial bonds involve sp^2 hybrids on the phosphorus.

Again it should be emphasized that the sets of equivalent hybrid orbitals are only strictly appropriate when the molecule is symmetrical. An infinity of less symmetrical sets of hybrid orbitals can be derived from other combinations of atomic orbitals. Thus, for example, we use the equivalent sp^3 hybrids for the methane molecule where the bonds radiate from the carbon nucleus in regular tetrahedral directions. However, for the case of ammonia, NH_3, we imagine the nitrogen to use a different set of approximate tetrahedral hybrids—the first three concerned with the NH bonds being equivalent and with interbond angles of 106.6°, and the fourth not being equivalent to the other three and accommodating the nitrogen lone-pair electrons.

In all cases the hybrid orbitals represent suitable combinations of the appropriate atomic orbitals of the valence shell. Such orbitals have similar orbital energies, which is a primary prerequisite for the formation of hybrids. Effective combinations are not possible when the atomic orbitals do not have similar energies.

The preceding description of the concept of hybridization makes it seem that it is "being wise after the event" to start from the known geometry of the molecule and then decide which hybrid orbitals are involved in bonding. However, if we can be sure of the nature of the valence orbitals that are to be mixed to give the hybrids, a prediction of molecular geometry is possible. In almost all cases the predicted geometry is the same as is predicted by the Sidgwick–Powell rule. However, there is one case in which the nature of the valence orbitals predicts a geome-

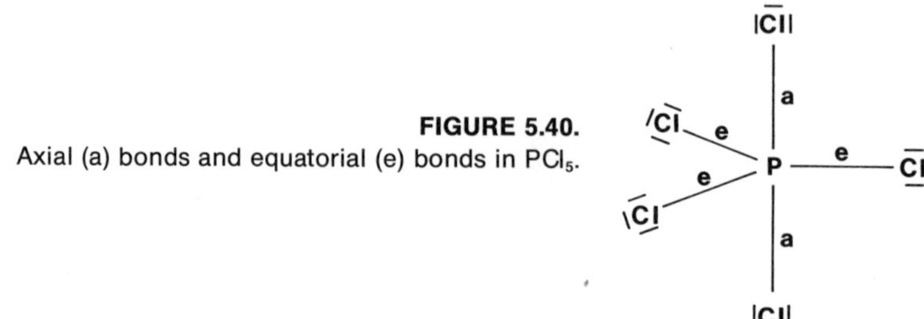

FIGURE 5.40.
Axial (a) bonds and equatorial (e) bonds in PCl_5.

try different from that derived by considering electron repulsion. It is the case of some compounds of Ni(II), Pd(II), and Pt(II), where the valence configuration is dsp^2. In this case the four hybrids all lie in one plane and so square–planar geometry is predicted for the four bonds radiating from the metal atom. Electron repulsion alone favors the alternative tetrahedral arrangement of bonds. It has been found experimentally that most compounds of nickel, palladium, and platinum in the +2 oxidation state do in fact have a planar geometry—a triumph for the hybridization concept!

Delocalized Molecular Orbitals

The idea of localized molecular orbitals formed from combining hybrid atomic orbitals on one atom with some atomic (hybrid or otherwise) orbitals on an adjacent atom is useful, as we have just seen for depicting molecular geometries. It is by no means a unique description, as we can equally well write molecular orbitals that are not localized between two centers but may spread over three or more atoms. These molecular orbitals are commonly known as *delocalized molecular orbitals*. We can illustrate how these delocalized molecular orbitals are formed with the example of a simple triatomic molecule called methylene, CH_2. Methylene is observed in hydrocarbon flames and is a stable molecule although extremely reactive. The coordinate system we choose as a framework for our description is shown in Figure 5.41, the z-axis being the internuclear axis. The electron configuration for the carbon atom in its valence state is

$$\text{C: K}(2s)^1(2p_x)^1(2p_y)^1(2p_z)^1.$$

We use the 2s and $2p_z$ atomic orbitals of carbon along with the 1s orbitals of the two hydrogen atoms H_A and H_B to form our molecular orbitals. The $2p_x$ and $2p_y$ atomic orbitals are left alone on the carbon atom containing one electron each. The appropriate linear combinations to form the molecular orbitals should concentrate electron density in the bond regions between the nuclei. There are two ways of doing this with the orbitals that are available.

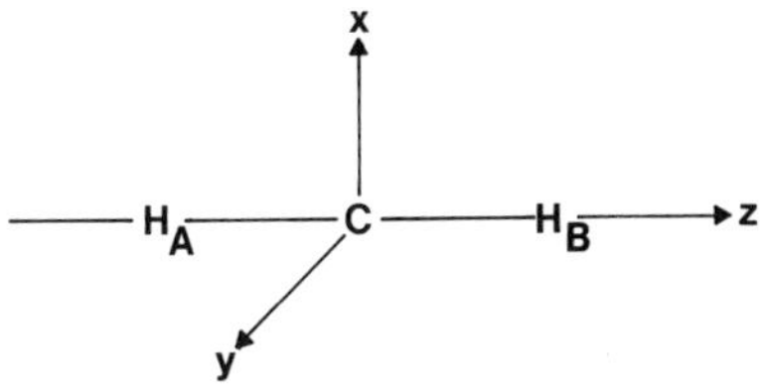

FIGURE 5.41. Coordinate system for CH_2.

$$\begin{aligned}\sigma_1 &= c_1 2s + c_2(1s_A + 1s_B)\\ \sigma_2 &= c_3 2p_z + c_4(-1s_A + 1s_B)\end{aligned} \tag{5.15}$$

The coefficients c_1, c_2, are mixing parameters that allow us to select how much of each orbital to include in the combination. These combinations give the molecular orbitals depicted in Figure 5.42. They are bonding molecular orbitals. As with diatomic molecules there are also antibonding molecular orbitals formed from the combinations:

$$\begin{aligned}\sigma_1^* &= c'_1 2s - c'_2(1s_A + 1s_B)\\ \sigma_2^* &= c'_3 2p_z - c'_4(-1s_A + 1s_B)\end{aligned} \tag{5.16}$$

These are depicted in Figure 5.43. Notice that given the algebraic expressions for the molecular orbitals, for example, Equations 5.15 and 5.16, the contour diagrams representing the orbitals can be sketched simply by considering the signs of the various atomic orbitals as they combine to form molecular orbitals. Also note that we always form the same number of molecular orbitals as we had atomic orbitals from which to form them (in this case, four). Notice further that σ^*-molecular orbitals always have more nodes than the corresponding σ-molecular orbitals.

The only remaining atomic orbitals in the valence shells of the atoms concerned are the $2p_x$ and $2p_y$ atomic orbitals on the carbon atom. These do not form molecular orbitals because the hydrogen 1s orbitals are of the wrong symmetry.

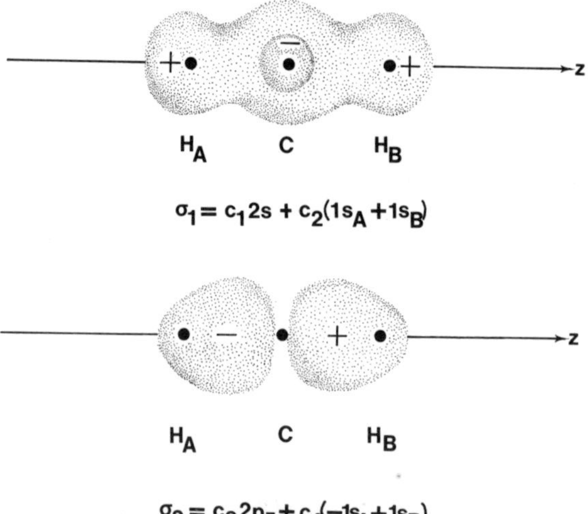

FIGURE 5.42. Bonding molecular orbitals for CH_2. The coefficients c_1, c_2, c_3, and c_4 are not specified but will not change the general appearance given.

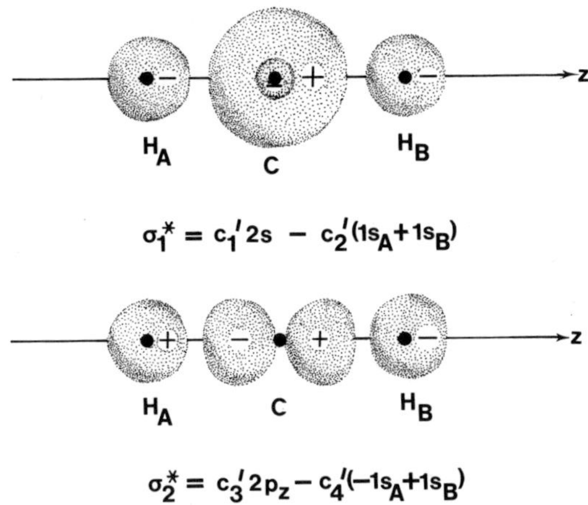

FIGURE 5.43. σ^*-Antibonding molecular orbitals for CH_2.

For our description of the electronic structure of CH_2 we need only know the filling order of our orbitals (see Figure 5.44). The total of six electrons (four from carbon and two from the hydrogen atoms) are fed in according to the Aufbau principle and Hund's rule to give the configuration

FIGURE 5.44. Filling-order diagram for CH_2 molecular orbitals.

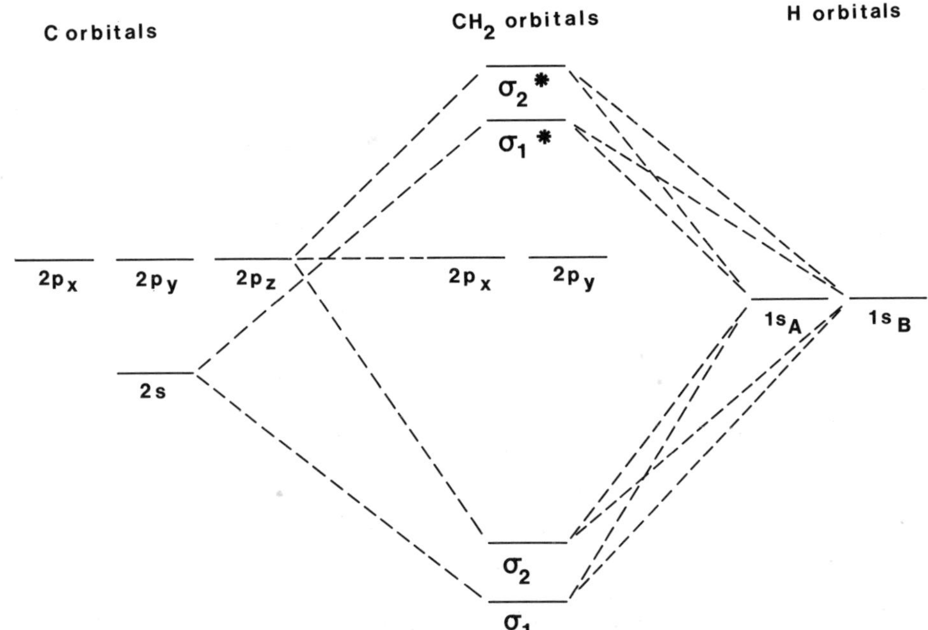

$$CH_2\text{: } (\sigma_1)^2(\sigma_2)^2(2p_x)^1(2p_y)^1$$

so that in the description of the electronic structure of CH_2 we have two σ-type electron-pair bonds delocalized over all three atoms and two unpaired electrons localized on the carbon atom. The observed paramagnetic behavior of methylene strongly supports this picture. The molecular orbital description with two unpaired electrons predicts the linear geometry that has been experimentally observed. The unpaired electrons also account for the high chemical reactivity of CH_2.

We note in passing that compounds such as CH_2 are *electron deficient* in the sense that they do not complete a valence shell (in this case an octet) around the carbon atom. Such compounds are difficult to depict using Lewis structures. The compounds are inevitably reactive—so much so, that they can usually not be kept in a reagent bottle. Some, though, are stable enough to be kept in glass containers, an example being boron trimethyl:

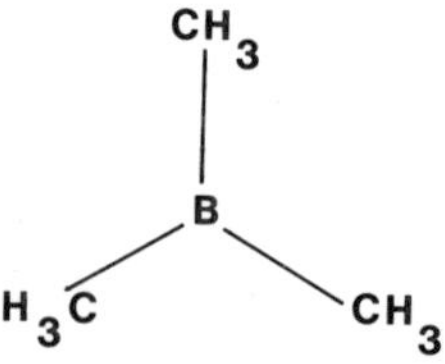

Many boron compounds with all valence electrons paired are electron deficient yet possess enough stability to be handled with caution. One compound whose electronic structure was a subject of controversy for many years is diborane[14] B_2H_6. It is now recognized that the two borons are joined by three-center bonds that include two of the hydrogens. It is sometimes said that the borons are united by a "protonated double bond," but it is easier to think of two delocalized molecular orbitals, each extending from boron to hydrogen to boron, as depicted in Figure 5.45, and each filled with an electron pair. Other larger boron hybrides are known and in all of them there are extended molecular orbitals embracing more than two atoms. The bonding description of electron deficient compounds using

[14]Diborane exemplifies a more restricted category of compound that has been referred to as *electron deficient*—one in which there are insufficient valence electrons to form the minimum number of electron pair bonds between pairs of atoms and to join all of the atoms together. (The minimum number of such bonds to join n atoms is $n - 1$. Diborane, with eight atoms, contains only six pairs of valence electrons.)

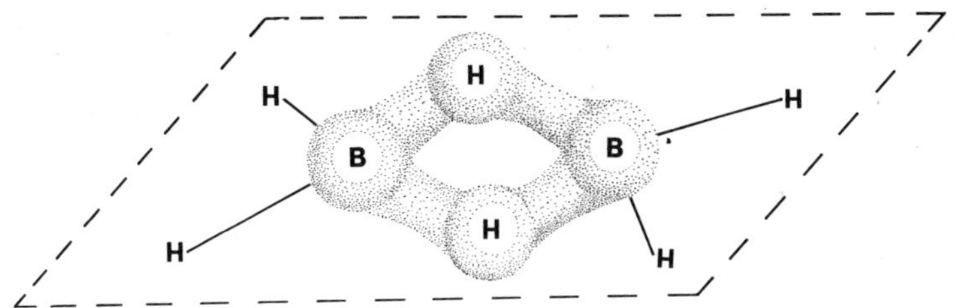

FIGURE 5.45. Three-center orbitals in diborane.

molecular-orbital theory presents no problems as we have seen for CH_2 (and H_2^+, e.g., and other first- and second-row diatomics). The *electron deficient* classification is rapidly becoming obsolete except when discussing chemical reactivites.

σ- and π-Molecular Orbitals in Polyatomic Molecules

Earlier where we described the electronic structure of diatomic molecules in terms of σ- and π-type molecular orbitals (pp. 101–113), you should have noticed that in molecules such as N_2 we could attribute the bonding to the set of electrons occupying the $2p\sigma$ and $2p\pi$ molecular orbitals in a $\sigma^2\,\pi^4$ configuration. This implies that the triple bond in N_2 consist of a σ-bond and two π-bonds.

This description is quite acceptable because the principles of quantum mechanics allow us to dissect the total electron distribution in whatever way is most helpful to describe the properties of the molecule. It is rather like cutting up a piece of cake. Sometimes we slice it into pieces of similar shape (equivalent bonds such as representing multiple bonds with extra dashes) or, on other occasions, we might choose to slice the cake into differently shaped pieces (σ- and π-bonds). We now want to consider briefly how we can use this second approach for polyatomic molecules with multiple bonds.

When a polyatomic molecule has a planar geometry (e.g., ethylene, butadiene, benzene, and sulfur dioxide, see Figure 5.46), the electron distribution in the portion of the molecule containing a double bond can be separated into σ- and π-components having shapes (viewed side-on) as shown in Figure 5.47. Thus, for example, in ethylene since there are three atoms around each carbon atom we can use sp^2 hybrid atomic orbitals and form localized σ-molecular orbitals to represent the four CH bonds and one CC bond as in Figure 5.48. The σ-molecular orbitals are each doubly occupied and hence account for 10 of the 12 valence electrons in C_2H_4. Each carbon atom has one 2p orbital that is not involved in the σ orbitals. These 2p

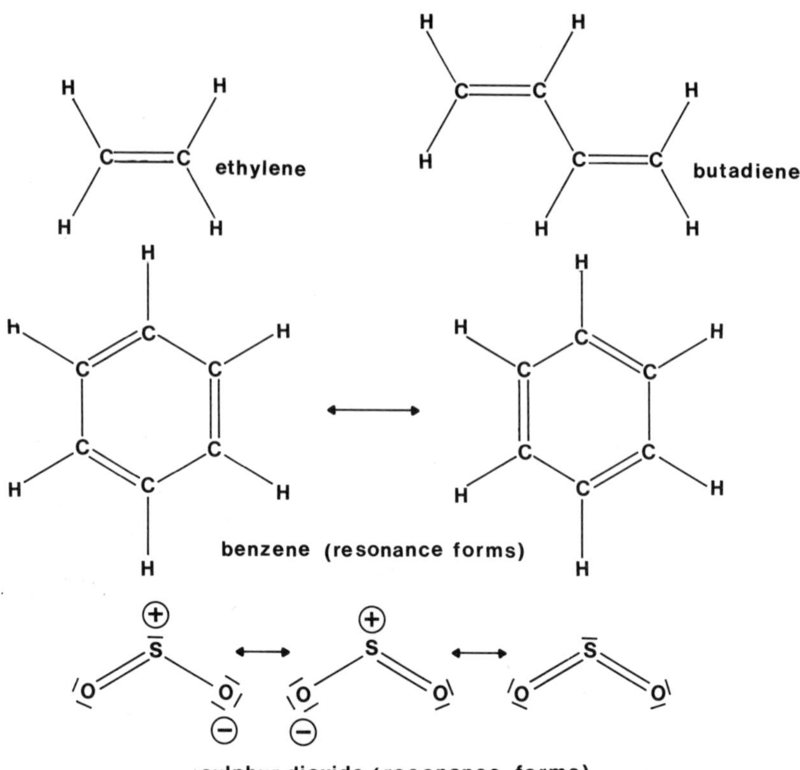

FIGURE 5.46. Some planar molecules containing double bonds.

orbitals are perpendicular to the σ framework (i.e., to the sp^2 hybrids) and combine to form a π-molecular orbital above and below the plane as in Figure 5.48. The π-molecular orbital is occupied by the remaining two valence electrons and together with the σ bond it accounts for the C=C double bond in ethylene. The total bonding of the ethylene molecule is represented in Figure 5.49.

In the case of molecules such as butadiene or benzene, where a chain or ring of atoms with alternating double and

FIGURE 5.47. Electron distribution of a double bond separated into σ- and π-components (viewed side-on).

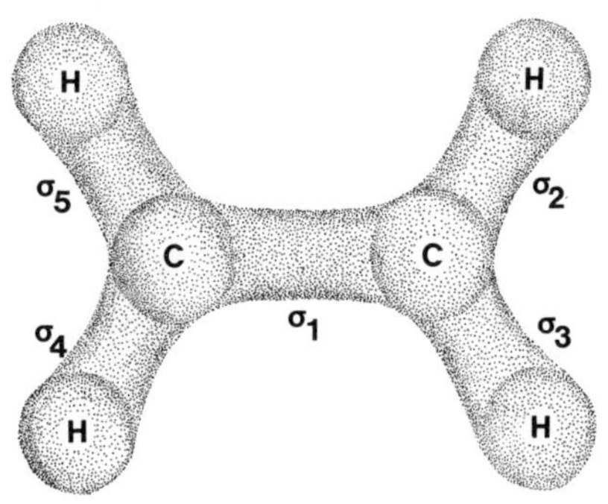

FIGURE 5.48. The σ-bond framework for ethylene, C_2H_4; σ_1 is formed from two sp^2 hybrids, one from each C atom. σ_2, σ_3, σ_4, σ_5 are MO's formed from an sp^2 hybrid on the carbon atom and the hydrogen 1s orbitals.

single bonds occurs, we can regard the π-molecular orbitals as extending over a number of atomic nuclei instead of just two nuclei as in ethylene. The delocalized π-molecular orbitals in these instances are formed from various linear combinations of the carbon 2p atomic orbitals directed perpendicular to the molecular plane. The rest of the framework of the molecule can again be thought of in terms of localized σ-molecular orbitals formed from sp^2 hybrids on the carbon atoms and hydrogen 1s orbitals. Taking the specific case of benzene, we can form six π-molecular orbitals extending over the entire benzene ring, the first three of which contain the six 2p electrons, one from each carbon atom (see Figure 5.50). The delocalization of the π-electrons makes all of the CC bonds of benzene equivalent. They are neither single nor double bonds but somewhere in between. Measurement of the CC bond length gives a value of 139.7 pm, confirming this picture because this value is intermediate between C—C (154 pm) and C═C (134 pm).

There is abundant experimental evidence that π-molecular orbitals have a greater orbital energy than σ-molecu-

FIGURE 5.49. Bonding scheme in ethylene highlighting the π-molecular orbital formed between the two carbon atoms. The σ-molecular orbitals are shown schematically.

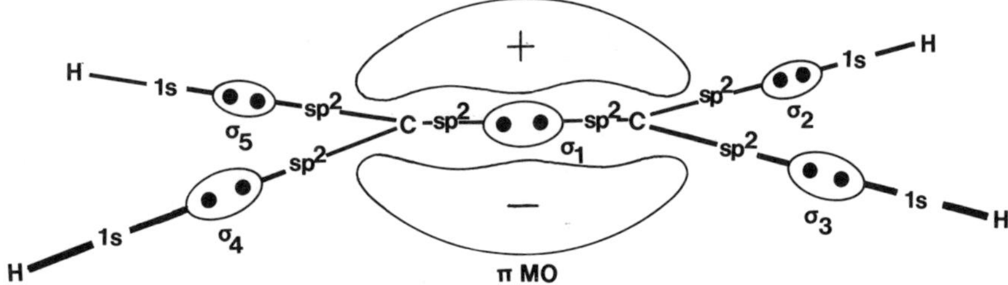

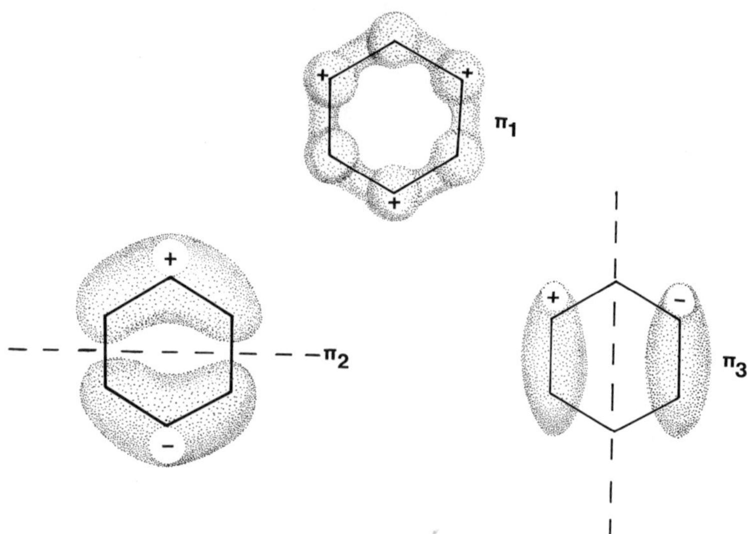

FIGURE 5.50. The three occupied π-molecular orbitals in benzene (viewed from above).

lar orbitals for the type of molecules we have been dealing with in this section. The evidence comes primarily from absorption spectra, ionization potentials, and chemical reactivities. The electron configuration of benzene with its 30 valence electrons (four from each carbon and one from each hydrogen) can be represented by Figure 5.51.

The molecular-orbital description of benzene directly emphasizes the way in which electrons can become delocalized and circumvents the problem of having to invoke the concept of the resonance forms in Figure 5.46 to account for the electronic structure of the benzene molecule.

A final example of a molecule where we can describe the bonding in terms of localized σ-molecular orbitals and

FIGURE 5.51. Orbitals in benzene used to accommodate the 30 valence electrons.

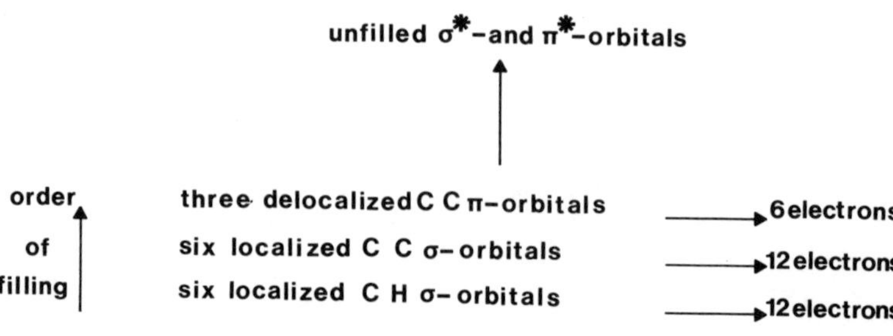

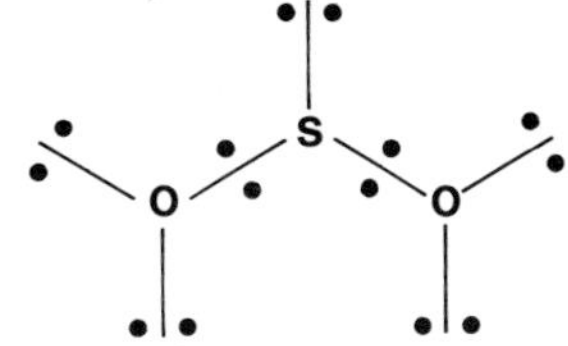

FIGURE 5.52.
σ-framework of SO_2. σ-Molecular orbitals are formed between the S and O atoms each containing two electrons. Other electrons occupy sp^2 hybrids as lone pairs. Orbitals are shown as lines for simplicity.

delocalized π-molecular orbitals is sulfur dioxide. Here there are a total of 18 valence electrons. We form sp^2 hybrids on each atom and then σ-orbitals between the sulfur and each oxygen atom. The two remaining sp^2 hybrids on each oxygen atom and one remaining sp^2 hybrid on sulfur are lone-pair orbitals. This process is illustrated in Figure 5.52. A total of 14 valence electrons is used in this process. The remaining four valence electrons will occupy delocalized π-molecular orbitals formed from 2p orbitals perpendicular to the sp^2 hybrid orbitals. We can form three molecular orbitals from the three 2p orbitals. One is a bonding molecular orbital, one is nonbonding (lone pair), and one is antibonding. The shapes are depicted in Figure 5.53. In the ground state of SO_2 the bonding and nonbonding π-molecular orbitals are occupied by the four remaining valence electrons.

We have been becoming increasingly qualitative in our molecular-orbital description of polyatomic molecules.

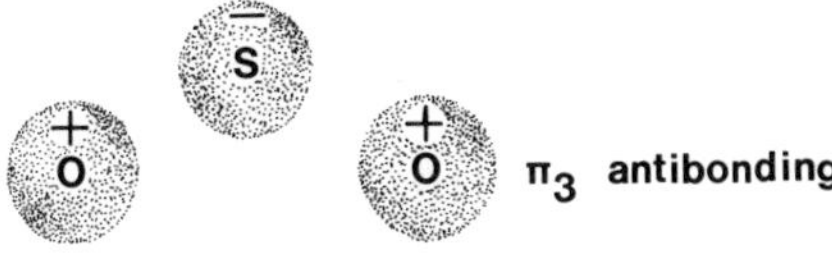

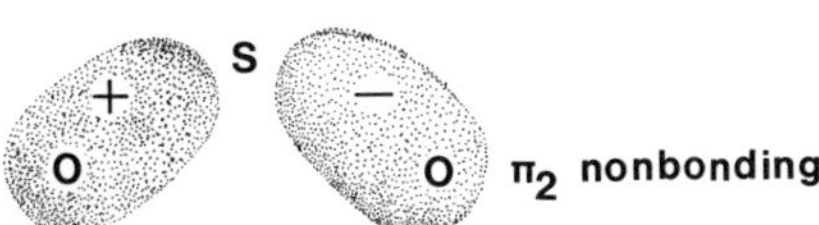

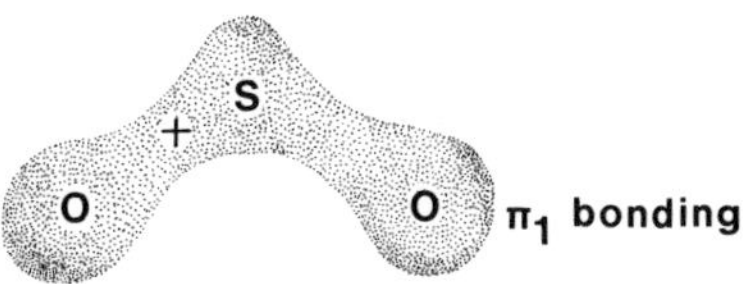

FIGURE 5.53.
π-Molecular orbitals in SO_2 (viewed from above). π_1 and π_2 are occupied in SO_2.

This is because the mathematical detail gets more and more complex. However, with modern high-speed computers there is no need for qualitative pictures such as dividing the electrons into sp-type hybrid orbitals, and thence into σ- and π-type molecular orbitals. We can solve the problems quite generally and arrive at similar answers because the total electron distribution is quite independent of the way in which we decide to dissect it for analysis.

Crystal-field Theory

The theories described in previous sections serve as a basis for discussion of many aspects of the electronic structures of molecules. However there is another theory that helps us to understand the magnetic and spectroscopic properties of some compounds—those of the transition elements, the lanthanides and actinides. We now give a brief introduction to this method, which is called *crystal-field theory*. It considers the effect on the energies of atomic orbitals of an atom when placed in the field of a set of negative charges, or an electrostatic field. In Chapters 3 and 4 we found that, in an isolated atom, the atomic orbitals occur in degenerate groups of three p-orbitals, five d-orbitals, seven f-orbitals, and so on. However, the energies of all atomic orbitals are altered if the atom is placed in an electrostatic field, and often the energies of the orbitals composing a degenerate set are not all altered by the same amount. For example, suppose that two negative charges are brought up to positions equidistant on either side of an atom on the z-axis (see Figure 5.54). Because of the charges, an electron has a higher energy in the p_z orbital, which is located in the neighborhood of the z-axis, than an electron in the p_x or p_y orbital, which have equally favorable positions relative to the charges. A splitting of the energies of the orbitals results. (see Figure 5.54).

From the various orientations of the d-orbitals relative to the x-, y-, and z-axes (Figure 3.21), the same qualitative electrostatic argument shows that the splitting given in Figure 5.54 will be produced by the field of the two negative charges.

The effects of fields due to four negative charges in a square planar arrangement and six negative charges in an octahedral arrangement may be similarly deduced by considering the positions of the various orbitals relative to the charges (see Figures 5.55 and 5.56). The difference in energy between the lower three and upper two d orbitals produced by the octahedral crystal field is denoted by Δ_{oct}.

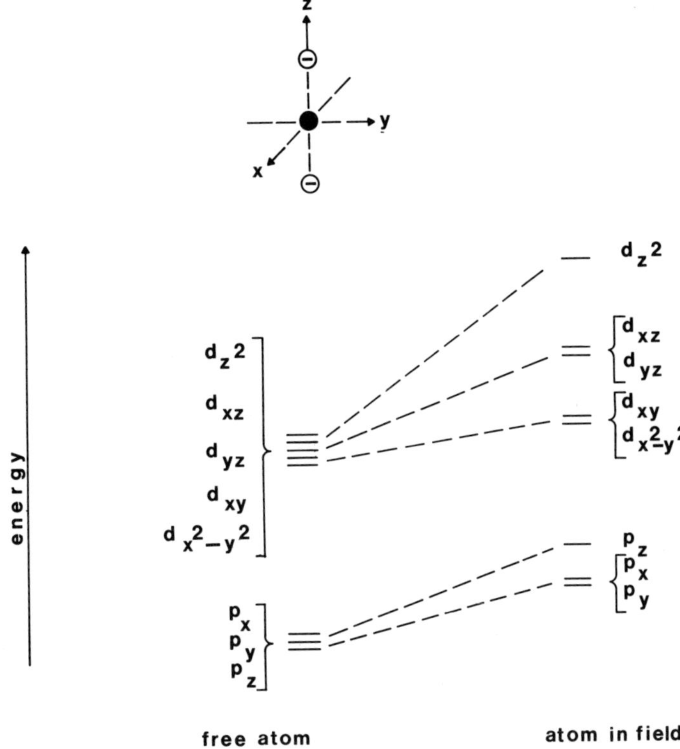

FIGURE 5.54. Atom in diagonal electrostatic field. Bracketed orbitals are degenerate.

The effect of four negative charges at the corners of a regular tetrahedron is slightly more difficult to visualize (see Figure 5.57). Notice that the splitting Δ_{tetra} of the degenerate d-orbitals is in the reverse direction to that caused by an octahedral field. A quantum-mechanical calculation shows that when the charges are the same distance from the central atom the tetrahedral field from four charges produces only four-ninths of the splitting caused by the octahedral field of six charges (see Figure 5.57).

The arrangements of negative charges that we have considered turn out to be those of most interest. If the fields due to positive charges are considered then the energy changes are just the reverse of those set out in Figures 5.54–5.57. However, we have concentrated on negative charges because the prime application of crystal-field theory arises when we take the extreme ionic view of the structure of a transition-element compound by considering it to be a central cation surrounded by a group of

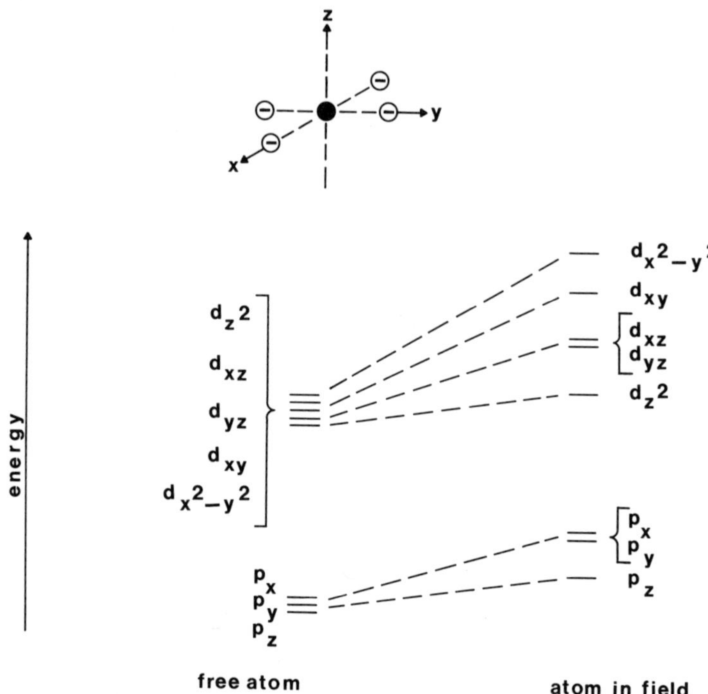

FIGURE 5.55. Atom in square–planar field. Bracketed orbitals are degenerate.

anions. We then use the principle summarized in Figures 5.54–5.57 to decide how the electronic structure of the central ion is affected by the field of the surrounding anions. In some cases this ionic viewpoint is not too extreme, for example, when we consider the Ti_2O_3 lattice as Ti^{3+} cations surrounded octahedrally by O^{2-} anions. But in other cases, as in the $Cr(H_2O)_6^{3+}$ cation where the six water molecules surrounding the chromium cation have to be regarded as equivalent to six negative charges, we are adopting an exaggerated ionic view of the structure. Crystal-field theory finds its main application with transition-element compounds where incompletely filled d-orbitals are often encountered. For the compounds of main group elements the d-orbitals are either completely filled or empty, and it does not prove fruitful to consider in detail the splitting of the p-orbitals since they are usually filled with valence electrons. The discussion of rare-earth and actinide element compounds from the point of view of crystal-field theory requires us to consider the splittings for f-orbitals. Since we have not presented their angular dependence, we do not discuss this. It follows in the same way as for the simpler orbitals.

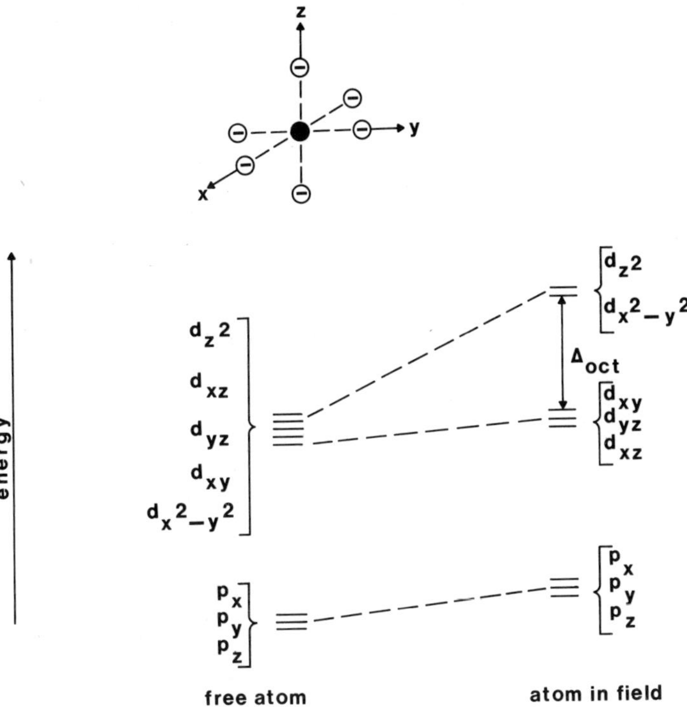

FIGURE 5.56. Atom in octahedral field. Bracketed orbitals are degenerate.

FIGURE 5.57. Atom in tetrahedral field.

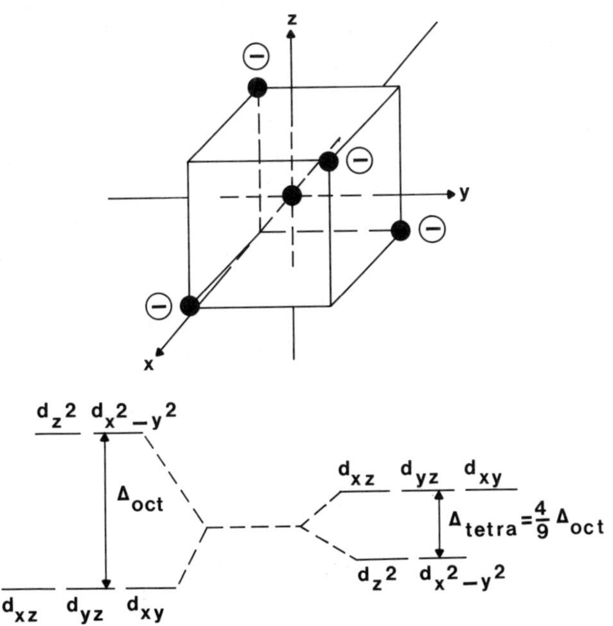

Crystal Fields Spectra, and Magnetic Properties

When the d-orbitals of an atom or ion are split by a crystal field into groups of differing energy and the d-orbitals are not completely filled, the possibility arises of an electron in the lower group of d-orbitals being moved into the group of higher energy by absorbing light of the appropriate wavelength. For example, let us consider an ion with a single d electron on its own and subjected to the octahedral field of six surrounding anions. The ion Ti^{3+} has a normal electronic configuration of $(1s)^2(2s)^2(2p)^6(3s)^2(3p)^6(3d)^1$. By absorption of light of the appropriate wavelength the electron in the 3d orbital may be excited to the 4p orbital (see Figure 5.58); the dotted arrow implies that the transition to the 4s orbital is not permitted according to the selection rules for spectra. However, in the presence of the octahedral field the 3d electron occupies one of the lower-energy d-orbitals. By absorption of light of energy corresponding to Δ_{oct} the electron is excited to one of the higher energy d orbitals. Thus by determining the absorption spectrum of the ion in a particular octahedral field we can discover the extent to which the d orbitals are split by the field. For example, the absorption spectrum of $Ti(H_2O)_6^{3+}$, which accounts for its purple color, is shown in Figure 5.59. The position of maximum light absorption corresponds to a wavelength of 500 nm or 20000 cm^{-1}, so that Δ_{oct} = 20000 cm^{-1} = 2.5 eV.

In a great many transition-element compounds the crystal-field splitting of the d-orbitals is such that an electronic transition from one set of d-orbitals to the other is produced by light absorption in the visible region of the spectrum (i.e., wavelengths in the range 400–700 nm corresponding to photon energies of 3.1 eV to 1.7 eV). Typical values of Δ_{oct} for some transition element cations are given in Table 5.11. For an ion of given charge the value of the splitting is roughly independent of the nature of the ion

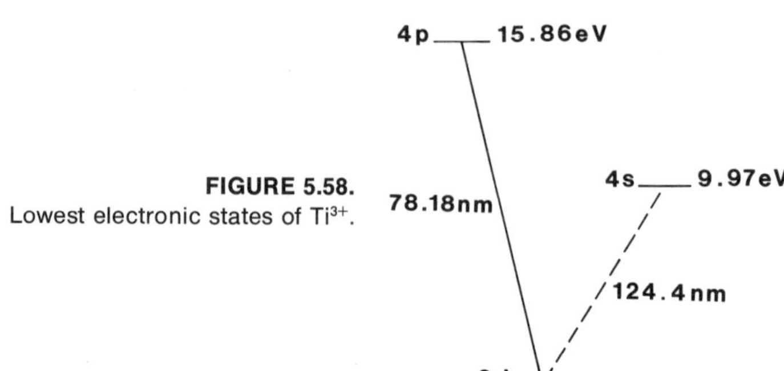

FIGURE 5.58.
Lowest electronic states of Ti^{3+}.

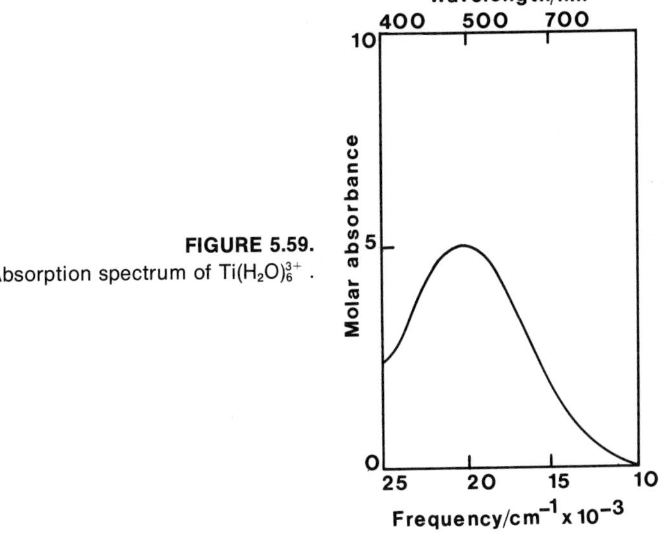

FIGURE 5.59. Absorption spectrum of $Ti(H_2O)_6^{3+}$.

but depends noticeably on the nature of the ligands creating the octahedral crystal field.

Such light absorption means that the compound is colored and so we can associate the fact that many compounds of transition elements are colored with the effect of crystal fields on d-orbitals of the atom. Notice in Figure 5.58 that the corresponding lowest-energy free-atom transition for Ti^{3+} is at 78.18 nm (in the far-vacuum ultraviolet).

Let us consider how the magnetic properties of a transition-element compound may be affected by crystal-field splitting. Consider the case of an ion with a d^6 structure,

Table 5.11 Some crystal-field splittings, Δ_{oct} (eV)

		Ligand		
		H_2O	NH_3	CN^-
Ti^{3+}	d^1	2.5		
V^{3+}	d^2	2.2		
Cr^{3+}	d^3	2.2	2.7	3.3
Mn^{3+}	d^4	2.6		
Fe^{3+}	d^5	1.7		
Co^{3+}	d^6	2.3	2.8	4.2
Mn^{2+}	d^5	1.0		
Fe^{2+}	d^6	1.3		4.1
Co^{2+}	d^7	1.2	1.3	

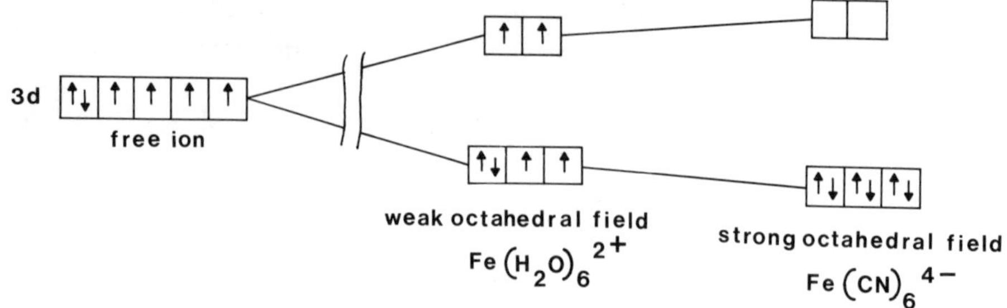

FIGURE 5.60. Electronic structure of Fe^{2+} compounds.

such as Fe^{2+} (Figure 5.60). The free ion has four unpaired electrons in the 3d orbitals and so is strongly paramagnetic. The same is true in the presence of a weak octahedral field where the splitting is so small that the orbitals are nearly degenerate for the purpose of Hund's rule. In a strong field, however, the splitting is sufficient for the six electrons to be confined to the lower-energy orbitals.

The decision between the strong and weak field cases depends on whether the energy P required to pair off two 3d electrons is greater than or less than the energy Δ_{oct} required to raise one of them into the upper orbitals. The energy difference between the two configurations shown in Figure 5.60 is $2P - 2\Delta_{oct}$. If this quantity is positive the first one is more stable; if it is negative, the second one is more stable. That is, ligands for which Δ_{oct} is small will produce the weak field structure, and ligands for which Δ_{oct} is sufficiently large will produce the strong field structure. The value of P can often be estimated from an analysis of the spectrum of the free atomic ion. For Fe^{2+} it is estimated to be 2.2 eV. The values of Δ_{oct} for various ligands can be derived from studies of absorption spectra, as illustrated in the preceding section. The data for Fe^{2+} are presented in Table 5.12.

The magnetic properties of compounds containing various numbers of d electrons and for fields of other symmetries can be discussed in a similar way. In all cases the

Table 5.12 Crystal-field data for Fe^{2+} (P = 2.2 eV)

Ligands	Δ_{oct} [eV]	$2P - 2\Delta_{oct}$ [eV]	Predicted	Observed
$6H_2O$	1.3	+ 1.8	Paramagnetic	Paramagnetic
$6CN^-$	4.1	− 3.8	Diamagnetic	Diamagnetic

decision between the weak-field and strong-field configurations depends on the relative values of P and Δ.

Van der Waals Forces

There are still attractive forces between atoms with filled valence shells (i.e., the rare gases) and between molecules when all of the valence electrons are used in completing the valence shells of the atoms in these molecules. These forces are much weaker than those involved in covalent bonds and are called *van der Waals forces* or sometimes van der Waals bonds. There are two sorts of forces involved, one repulsive and one attractive. The repulsive force merely involves the repulsion between electrons in filled orbitals on neighboring atoms or molecules. The attractive force arises when instantaneous dipoles on neighboring atoms or molecules attract each other.

Instantaneous dipole moments need further explanation. To do this we take as an example the atoms of a rare gas. The electrons in the atom are continually in motion. Thus the rare-gas atom, which on the average has a spherically symmetric charge distribution, will at some instant in time have a nonspherical charge distribution, which gives it a resultant dipole. This instantaneous dipole can induce a dipole moment in a neighboring atom, oriented such that an attractive force is experienced. The two forces give rise to a potential-energy versus distance-of-approach curve quite like that in Figure 5.1(c), except that the distance at which the minimum occurs is much greater, and the curve generally has a shallower minimum than that for the potential-energy curve for a covalent bond. A comparison of a potential-energy curve for the van der Waals force between helium atoms and the covalent bond in H_2 is given in Figure 5.61. The covalent bond is nearly 6000 times more stable than the van der Waals bond. The other notable feature about van der Waals bonds is that they increase in strength as the size of the atom increases. This is presumably because for larger atoms the electrons are held less strongly, thus causing larger instantaneous and induced dipoles to occur. This increase in strength is reflected in the boiling points of the rare gases given in Table 5.13.

For molecules with both symmetrical and unsymmetrical structures the same type of force occurs. However, for molecules with permanent dipole moments there are additional intermolecular forces arising from dipole–dipole forces (from the permanent dipole) and from dipole-induced dipole forces (the former again referring to the

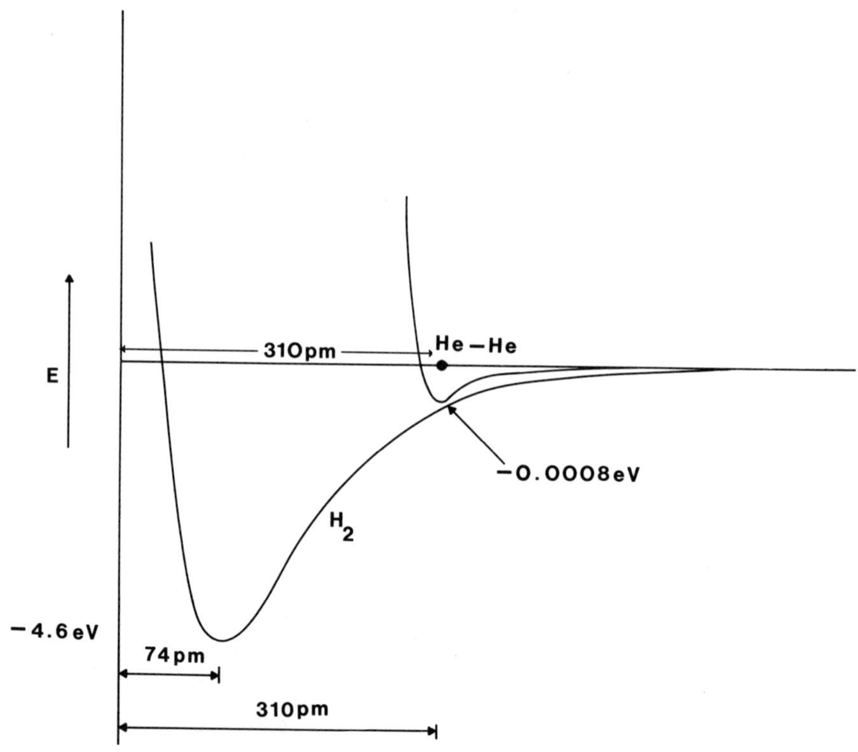

Figure 5.61. Comparison of approximate potential-energy curves for H_2 and He—He.

permanent dipole). Table 5.14 shows the relative contribution of these different types of force to the total intermolecular force for various molecules. We see that except when the permanent dipole moment is very high, the van der Waals forces make the dominant contribution. Again,

Table 5.13 Boiling points of the rare gases

Rare gas	Boiling point [K]
He	4.2
Ne	27
Ar	87
Kr	121
Xe	164
Rn	211

Table 5.14 Contribution of different forces to total intermolecular force

Molecule	Dipole moment [Debye]	van der Waals energy[a]	Dipole–dipole energy[a]	Dipole-induced dipole energy[a]
HI	0.38	1600	1.5	7
HBr	0.78	736	26	17
HCl	1.03	440	78	23
NH_3	1.5	390	351	42
CO	0.12	282	0.015	0.24

[a]In units of 10^{-24} J for molecules $10a_0$ apart.

as for rare gas atoms, the melting and boiling points increase as the size of the molecules increases. This is partly a mass effect. But it also depends on the increased strength of van der Waals forces with increased molecular size. A familiar example of this occurs for the straight-chain hydrocarbons, C_nH_{2n+2}. For $n < 5$, we have natural gas; for $5 < n < 11$, petrol; for $12 < n < 16$, kerosene; and oils and waxes for higher values.

Hydrogen Bonds

Under special circumstances the dipole–dipole forces we have just discussed may be unusually large. This occurs whenever a hydrogen atom is attached to an electronegative atom such as fluorine, oxygen, nitrogen, or chlorine. The resulting positively charged hydrogen atom will form a bond to another small, electronegative atom, again usually F, O, N, or Cl because the small size of the hydrogen atom allows it to approach unusually close to the electronegative atom. These bonds arise as a result of the dipole–dipole interactions and account for the strong association of molecules such as water and hydrogen fluoride (see Figure 5.62). The interactions are of sufficient importance in chemistry to be given a special name; they are termed *hydrogen bonds*. The unusual variation of the

FIGURE 5.62. Association of hydrogen fluoride.

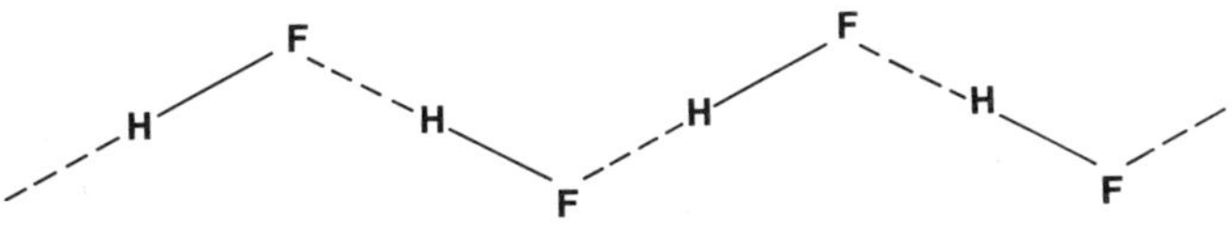

boiling points of hydrides of nonmetals shown in Figure 5.63 may be interpreted in terms of hydrogen bonds. The boiling point of the hydride of the smallest nonmetallic atom in groups V, VI, and VII of the periodic table is considerably higher than would have been anticipated from the boiling points of other members in the group. The high boiling point of the first member (i.e., HF, H_2O, and NH_3) is attributed to association of molecules in the liquid state by hydrogen bonds. Hydrogen bonding is less important for subsequent members in a given group because the larger atoms are less electronegative and hence do not form hydrogen bonds. Likewise CH_4, the first member of the hydrides of elements in group IV, does not have an abnormal boiling point because carbon is not sufficiently electronegative to form hydrogen bonds.

Hydrogen bonds play an important role in determining the properties of many compounds of biological importance. For example, the spiral structures of proteins and nucleic acids are held together by hydrogen bonds. In Chapter 6 we show further structures where hydrogen bonding is important.

Although hydrogen bonds are appreciably stronger than

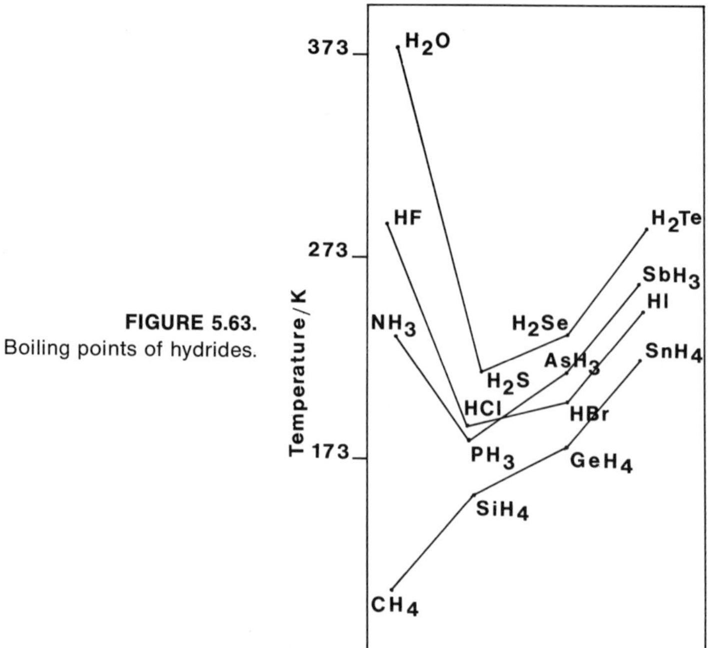

FIGURE 5.63. Boiling points of hydrides.

van der Waals bonds, they are still not nearly as strong as ionic or covalent bonds, a typical hydrogen bond energy being 10 kJ mol^{-1} compared to a covalent bond energy of around 500 kJ mol^{-1}.

Summary

We close this chapter by pointing out that although we have used many seemingly different theories to explain the bonding in various types of molecules, we have done so only for convenience. The quantum theory of electronic structure applied with sufficient rigor can adequately handle all types of molecules from simple diatomic molecules to massive biological macro-molecules.

Still, we have had an introduction into the realm of molecular theory and have seen how various descriptive theories can account for observed experimental parameters. These parameters include bond lengths and interbond angles (geometries of molecules), bond energies, dipole moments, and magnetic behavior. For simple ionic compounds an electrostatic theory based on Coulomb's law served our purposes well. For covalently bonded diatomics we introduced molecular-orbital theory and saw how, with adjustments due to electronegativity differences, molecular orbital theory could also account for observed characteristics of simple diatomic ionic molecules.

Molecular orbital theory is aesthetically pleasing because it harks back to what we learned about atomic orbitals, which in turn had a firm quantum mechanical foundation. It is a general theory, not just limited to diatomic molecules, but also applicable to triatomic molecules such as CH_2 and SO_2 as well as larger molecules such as benzene. We can form localized molecular orbitals to give a desired geometry by first hybridizing the atomic orbitals involved, or we can form delocalized molecular orbitals spread over the whole molecular framework. The total electron distribution of an atom or molecule is independent of how we picture it for convenience. We can make various molecular orbitals localized or delocalized, σ- or π-type by obeying certain quantum-mechanical rules, but always the total electron distribution is the same.

Sophisticated bonding theories (such as advanced molecular-orbital theory) necessarily require high-speed digital computers to carry out complex and repetitive mathematics. Most chemists (and other scientists, e.g., biologists and biochemists) are satisfied to use simpler bonding theories for polyatomic molecules such as those we discussed that were proposed by Lewis, Sidgwick, Powell, and others.

Problems

5.1 The normal internuclear distance in KBr is 329 pm. The ionization potential of potassium is 414 kJ mol^{-1}, and the electron affinity of bromine is 335 kJ mol^{-1}. Calculate the energy change for the formation of the ion-pair K^+Br^-.

5.2 Calculate the energy change for the formation of a dimer of Na^+Cl^- arranged together in a square:

$$\begin{matrix} Na^+ & Cl^- \\ Cl^- & Na^+ \end{matrix}$$

5.3 Write down in order of increasing energy the first 10 molecular orbitals used to describe the electronic structure of homonuclear diatomic molecules.

5.4 The O_2 molecule and its ions have the following experimentally observed O—O interatomic distances:

O_2^+	112 pm
O_2	121 pm
O_2^-	130 pm
O_2^{2-}	148 pm

Explain the increase in distance from O_2^+ to O_2^{2-} using simple molecular-orb1tal theory. Which of the above molecules would you expect to show paramagnetism, and why?

5.5 Give a molecular-orbital description of the LiF molecule. Compare this with a description of LiF using the electrostatic theory for combination of ions.

5.6 Write down electronic structural formulas for each of the following molecules. Your formulas should include the geometrical arrangement of atoms, formal charges, if any, and reasonable resonance structures.

a. SO_4^{2-}
b. O_3
c. CH_3COO^-
d. ClF_3
e. N_3^-
f. SF_4
g. POF_3
h. HNO_3
i. PO_4^{3-}
j. ClO_3^-

5.7 Draw three resonance structures for cyanogen azide, CN_4, given that the atoms are bonded in the order NCNNN. Indicate all valence electrons and formal charges. Can you predict whether the molecule will be bent or linear?

5.8 Which molecules from the following list should possess a permanent dipole moment?

a. $CHCl_3$
b. SO_4^{2-}
c. 1,3,5-trinitrobenzene
d. SO_2
e. CO_2
f. C_2H_2

5.9 Which hybrid orbitals might you use to describe the observed geometry of the following molecules?

a. ethane
b. SF_6
c. $Ni(CN)_4^{2-}$
d. acetylene
e. formaldehyde

Does the hybridization concept give a unique explanation of any of the above?

5.10 Deduce a delocalized molecular-orbital description of BeH_2 assuming a linear structure.

5.11 Formulate a molecular-orbital description for the ozone molecule, O_3. Compare this with the Lewis structure(s).

5.12 The ions $[Mn(H_2O)_6]^{2+}$ and $[Mn(CN)_6]^{4-}$ differ in color and magnetic properties.

a. Draw energy-level diagrams illustrating the differences in crystal-field splitting of the d-orbitals for the two ions and the distribution of electrons between the orbitals.
b. Why do the ions have different magnetic properties?
c. Which ion would you predict to absorb visible radiation of higher frequency?

5.13 A certain transition metal ion forms two octahedral complexes from two different ligands, one whose solutions is red and the other blue. Using simple crystal-field theory explain which complex would be expected to arise from ligands that create the stronger field.

5.14 Arrange the following molecules in order of increasing boiling points of their liquids: pentane, decane, and ethanol.

5.15 Why does xenon liquify at 166 K, whereas helium liquifies at 4 K?

5.16 Formic acid, HCOOH, forms a dimer in the vapor phase. Deduce a reasonable structure based on hydrogen bonds between the two formic-acid molecules.

6 The Solid State

In Chapter 5 we studied the structure of molecules and were careful to describe the spatial arrangement of the atoms in the molecule. In this chapter we do the same thing for the solid state, but here we may have chunks containing billions and billions of atoms, molecules, or ions. We are saved from this seemingly great complexity by the fact that the solids we study are atoms, molecules, or ions held in ordered arrays in space—this ordered arrangement is referred to as a *lattice* (a lattice is a regular periodic arrangement of points in space). We also treat the structure of perfect solids disregarding both imperfections that arise during growth of solids and deviations from the perfect structure at the surfaces where the forces are different from those in the middle of the crystal. Let us begin by describing the four main classifications of bonding in the solid state. They may be loosely called ionic, metallic, molecular, and covalent. Each of the four types give rise to distinctive properties that we discuss in some detail.

Ionic Solids

The building blocks of ionic crystals are the cations and anions formed when a metal and a nonmetal combine to form an ionic crystal. In Chapter 5 we discussed in some detail the forces involved in the bonding in an electrostatic model. Because of the very strong bonds between the ions throughout the crystal, they have high melting points and boiling points. Some examples are given in Table 6.1. Another general property of ionic solids is poor electrical conductivity (the electrons are tightly bound to each ion, whether cation or anion, and these are too large to move appreciably in the tightly packed structure).

Metals The metallic elements are those having one or only a few loosely bound electrons outside the outermost filled shell. It should be noted that the inner shells may be complete, for example, Na and Zn. On the other hand, some of the orbitals with $l > 2$ in inner shells may be only partially filled, for example, Fe or Pt.

It is relatively easy to give a description of the bonding and properties of metals by extending qualitatively our previous treatment of molecular orbitals for a triatomic system to the case when there are essentially an infinite (of the order of 10^{23}) number of contributing atomic orbitals. To be specific, let us consider the case of sodium, where for each atom there is one loosely bound electron (a 3s) outside a rare-gas configuration (Ne). The sodium ions in sodium metal are arranged in an infinite array (lattice, see later for details), and the 3s electrons hold them together in the following way. From Chapter 5 (pp. 101–104) we have seen that when two atomic orbitals combine to form molecular orbitals we get two, one higher in energy and one lower than the uncombined atomic orbitals. Similarly, when we have three atomic orbitals, we get three molecular orbitals. As the number of combining atomic orbitals increases a pattern emerges where from n atomic orbitals we get n molecular orbitals, some lower in energy and

Table 6.1 Properties of different types of solids

Compound	Bond type of crystal[a]	Melting point [K]	Electric conductance
Sodium chloride	Ionic Na^+Cl^-	Very high—1081	Solid—none; liquid—high
Magnesium chloride	Ionic $Mg^{2+}2Cl^-$	High—678	Solid—none; liquid—medium
Silicon tetrachloride	Molecular	Low—205	Solid—none; liquid—little
Chlorine	Molecular	Low—170	Solid—none; liquid—none
Benzene	Molecular	279	None
Sodium	Metallic	371	High
Diamond	3D-Covalent	Very high	None
Graphite	2D-Covalent + 1D-molecular	Very high	Slight and anisotropic
Copper II chloride	1D-Covalent + 2D-molecular	Very high	None
Plastic sulfur	Covalent + molecular	Low	None

[a] 1D, 2D, and 3D mean one, two, and three dimensional.

some higher than the original combining atomic orbitals. When the original combining atomic orbitals are of identical energy, we can see in a qualitative way that the total energy spread of the molecular orbitals formed from two or three will be much the same as for n. This is because physically the two can get close together, whereas for n some will be close and some far apart. We do not attempt to specify the orbitals here, but just illustrate the general energy features in Figure 6.1, since the three-dimensional nature of the metallic structure would introduce many complications. The important features for the description of metals are that: (a) the valence orbitals (one from each atom in the case of sodium) combine to form molecular orbitals that extend throughout the entire solid structure

FIGURE 6.1. Molecular-orbital (band theory) energy levels for valence electrons in metals.

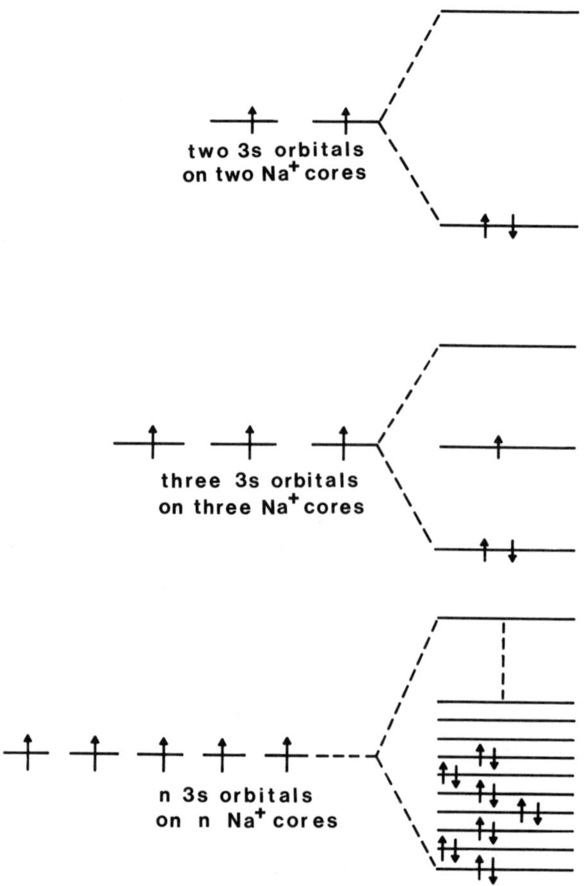

and have energies very close together and (b) the band of energy levels is only half filled. This last feature is, of course, explicable by the Pauli principle. For sodium, n atoms give n molecular orbitals and the n valence electrons pair up in the $(n/2)$ lowest-energy molecular orbitals. Thus the band of energy levels is only half filled.

The bonding in metals can then be described by saying that the valence electrons are in molecular orbitals that extend throughout the metal, binding the positively charged cores (for sodium, the Na^+) together. Where there are incompleted inner shells (d shells) these electrons may form additional directed bonds with neighbors (hard metals; e.g., Fe), whereas when the cores consist of rare-gas structures this is not possible so that the metals are softer (e.g., Na).

The main distinctive properties of metals are high electrical conductivity and thermal conductivity. Both of these properties have a simple explanation in terms of the band theory just given. When an electric field is applied to a metal, electrons can easily be excited to energy levels in which their momentum is such that they move in the direction of current flow. The energy gap between the levels in the band is very small. Since every level has a wave function extending over the entire solid, electron movement is very easy. This same feature also explains thermal conductivity. Heat applied at one position in a solid is easily transferred through the solid by electrons excited to levels just above the normally occupied ones.

Molecular Crystals

In molecular crystals the basic units are molecules rather than ions. The forces that hold the molecules together in the solid state are van der Waals forces and, if appropriate to the solid in question, hydrogen bonds (see Chapter 5). Because these involve much weaker forces than the forces holding ions together in ionic solids, it is much easier to overcome the forces holding the molecules in the ordered array in the solid state, so that they are characterized by fairly low melting points. They do not conduct electricity very well in either the solid or liquid state because the basic units of the crystal are uncharged.

Covalent Solids

The final classification we have chosen for solids can be further subdivided into three main types, but each is characterized by having covalent bonds that extend much further than those that hold the atoms together in normal-sized molecules. The first, which may be called *three-*

dimensional, has covalent bonds that bind the atoms together in all directions in the solid. Solids like this, such as diamond, are very hard and have extremely high melting points, since to melt them one has to disrupt essentially all of the covalent bonds in the solid. The second kind are those in which the covalent bonds are only in two directions, the bond in the third direction being either the very much weaker van der Waals attraction or some other kind of very much weaker bonding. These are often called *layer lattices* and have properties consistent with the very strong bonds in planar layers held together by very much weaker forces. Some types make good lubricants, because the planar layers can slide against one another fairly easily. The third type are those in which the covalent bonding extends only in one direction and the long strands are held together in the solid state by weaker forces as above. These are often referred to as *chain lattices*. One of the most outstanding properties is their fibrous nature. This results from the strongly bound chains being wrapped around one another with much weaker bonding forces.

Crystal Structures

Before proceeding with more detailed descriptions of the different properties of the different kinds of solid-state structures, we discuss the details of the structures. A convenient starting point is to look at the details of close packing of spheres. This is because the two different types of close packing of spheres are the structures that most metals adopt. Also the structures of many ionic solids can be viewed as having the anions (or cations) as a close-packed structure with the cations (or anions) occupying some of the sites or holes in between.

There are a few important concepts involved in the description of all crystals structures. They are as follows:

1. The *lattice,* which is an ordered array of *lattice points*. These are repeatable units of the crystal.
2. The *unit cell,* which is defined as the smallest volume that when repeated appropriately in three dimensions forms the complete crystal. It may be defined to be a number of different volumes in the crystal, each of which when repeated give the complete crystal. However, we use what is called the *conventional unit cell,* which is a small volume, but often not the smallest, that repeats itself in all directions to give the complete crystal. The reason for this is that often the conventional unit cell gives a much clearer idea of the packing

in the crystal than the unit cell. With the latter one also has to specify the directions and distances to understand the structure, whereas with the former a few of the repeat units (its volume is nearly always bigger) are already included.

3. The *number of molecules* or *ions of each type included in the unit cell.* The ratio of these will always give the empirical formula.
4. The *coordination number* of each different type of ion in the crystal. This is the number of closest spaced neighbors in the crystal. The polarity of neighbors (i.e., whether cation or anion) may be specified, but usually we are interested in the number of anions closest to a cation or the number of cations closest to an anion.

These concepts are illustrated as we describe the crystal structures.

Close-packed Structures

The positively charged atomic cores in metallic structures and the anions and cations in ionic crystals mostly are very close to spherical in shape so it is appropriate to begin by investigating the close packing of identical spheres.

If we lay a sphere on a table and try to put as many around it on the table as closely as possible we will find that the maximum number we can put is six. This is illustrated in Figure 6.2, where the circles represent the spheres, which we will call "a" spheres. Only the central one is marked in the figure.

If we continue to add spheres (all of equal size) in the plane of the table, we will find that each sphere has six equidistant touching neighbors. In the plane, the coordination number of each sphere is six.

To extend our model to a three-dimensional structure, if we look carefully at a model of seven spheres (as in Figure 6.2), we see that there are six indentations, labeled bcbcbc

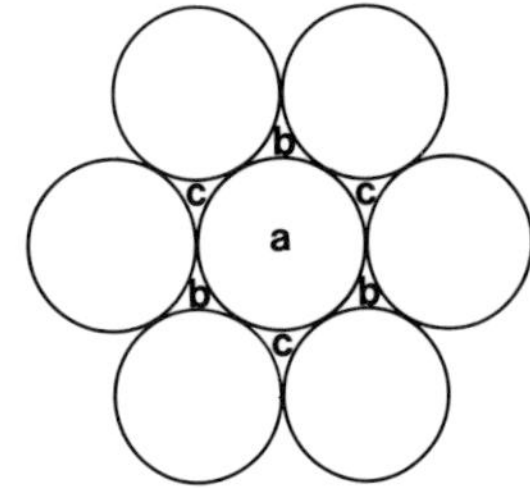

FIGURE 6.2. Illustrating the six nearest-neighbor spheres to a sphere in a plane.

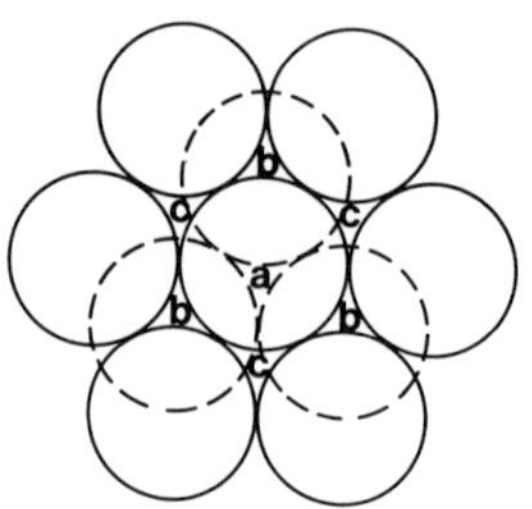

FIGURE 6.3.
Illustrating the second layer ("b" layer) in close-packed structures.

between the seven spheres. To place another sphere as close as possible to "a" and to start a layer as close above layer "a" (those on the table) as possible, we put the sphere in the position marked "b." When we have done this we find that there is not enough room to put other spheres in the adjacent "c" indentations, but that we can put two more in the other "b" ones. This means that we can rest a "b" sphere over every nonadjacent indentation in the "a" layer. This is the "b" layer as illustrated in Figure 6.3. As for the "a" layer, it extends across the entire plane. This is illustrated again in Figure 6.4 with a few more "b"-layer spheres added. It can be seen that within the "b"-layer plane the coordination number is again 6. Now it is also evident that the "b" layer sphere with the six closest neighbors in the plane has three "a"-layer spheres touching it as well. We see later that when the third layer is added, there will be three in that layer touching it. Thus the coordination number of a sphere in a close-packed structure in three dimensions will be 12.

Referring again to Figure 6.4, we can see that since the second layer is just like the first, the third layer of close-packed spheres will go into alternate depressions in the second layer. If we have chosen the "b" depressions for the second layer, we will have two choices for the third layer. They will be either directly above the "c" depressions in the first layer or directly above "a" spheres in the

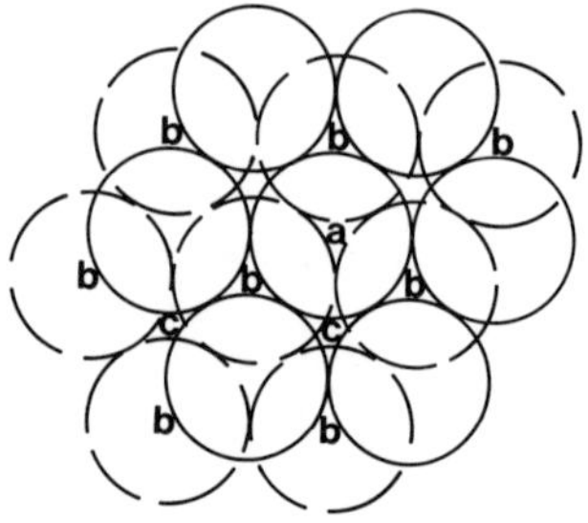

FIGURE 6.4.
Illustrating the second layer ("b" layer) in close-packed structures.

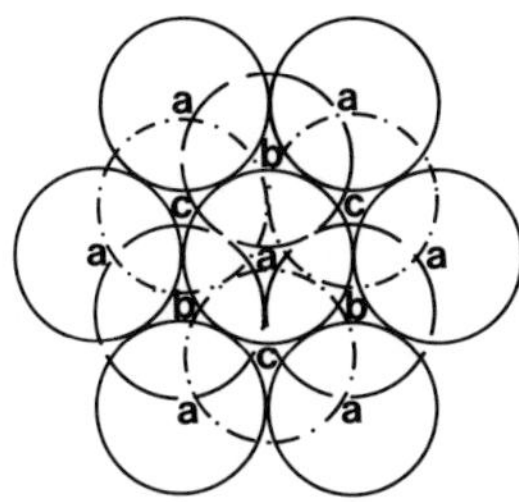

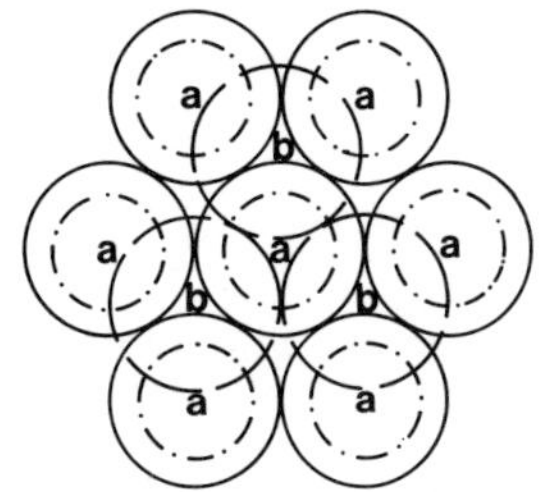

FIGURE 6.5. The repeatable layers of (a) CCP and (b) HCP. (In (b) the "c" layer lies directly over the "a" layer).

first layer. These two choices give two different close-packed structures. The first with an abc structure is called a cubic close-packed (or CCP) lattice and the second with an ab structure hexagonal close-packed (or HCP) lattice. The successive layers in them are respectively abcabc, and so on and ababab, and so on (see Figure 6.5). For both types of close packing the coordination number is 12. For each the amount of available space used is 74 percent; in other words, in the spaces between the spheres there is 26 percent of the total volume.

The Equivalence of CCP and FCC Unit Cell

The cubic close-packed structure that we have described above turns out to be the same as a very common unit cell called the face-centered cubic (FCC) structure illustrated in Figure 6.6. This structure has lattice points at the eight corners of a cube and others at the center of each of the six faces. Without three-dimensional models it is not easy to see the equivalence to cubic close-packed. However, the planes of spheres are easy to pick and are labeled by a, b,

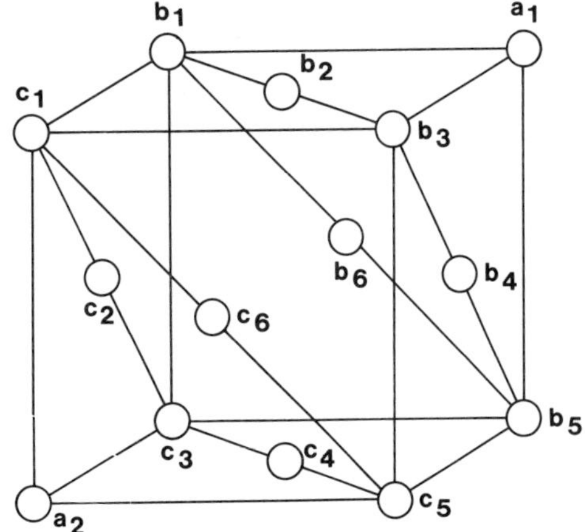

FIGURE 6.6. Face-centered cubic structure.

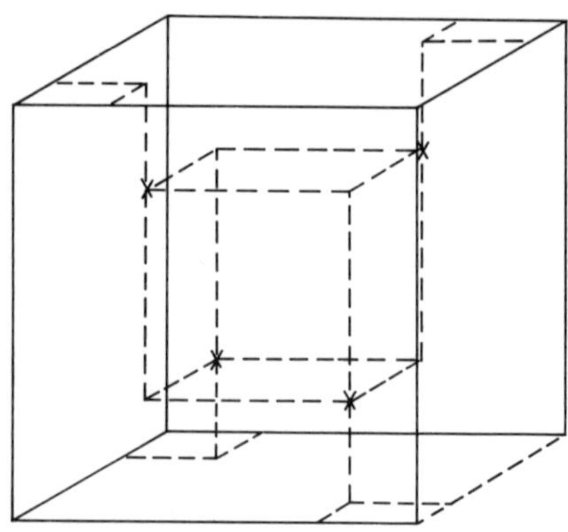

FIGURE 6.7.
Alternative FCC (or CCP) unit cell. The inner cube has sides half the length of those of the outer one and the lattice points are at the corners shown.

and c in Figure 6.6. With a little thought, one sees that a_1 lies on the $b_2b_4b_6$ depression, b_2 on $c_1c_2c_6$, b_4 on $c_4c_5c_6$, and b_6 on $c_3c_2c_4$. Then a_2 is one of the next "a" layer directly below a_1.

Figure 6.6 also represents what is the conventional unit cell for both cubic close-packed and face-centered cubic. However, one has to be careful in deciding the number of lattice points in it. When any entity of finite size is put at a lattice point it is shared by eight unit cells for the corner positions and by two unit cells at the face-centered positions. Since there are eight of the former and six of the latter, there are $8(1/8) + 6(1/2) = 4$ lattice points per unit cell. This is made somewhat clearer if we move the unit-cell position in the lattice one quarter of the length of the cube sides in three mutually perpendicular directions. Then, as shown in Figure 6.7, we get the unit cell of the same volume as the one shown in Figure 6.6, but clearly containing exactly four lattice points all within the volume.

We must decide on which unit cell to use. That shown in Figure 6.6 has the disadvantage that we have to dissect atoms or ions or whatever are at the lattice points to determine how many are in the unit cell. However, in more complex cases it is often impossible to avoid dividing molecular ions or molecules; hence we choose to retain the conventional unit cell as shown in Figure 6.6 as our standard way of representing it.

The Unit Cell of HCP

The conventional unit cell for the HCP structure is shown in Figure 6.8. Here there are $3 + 2(1/2) + 12(1/6) = 6$ lattice points per unit cell. The unit cell shown is three times the volume of the smallest repeating unit.

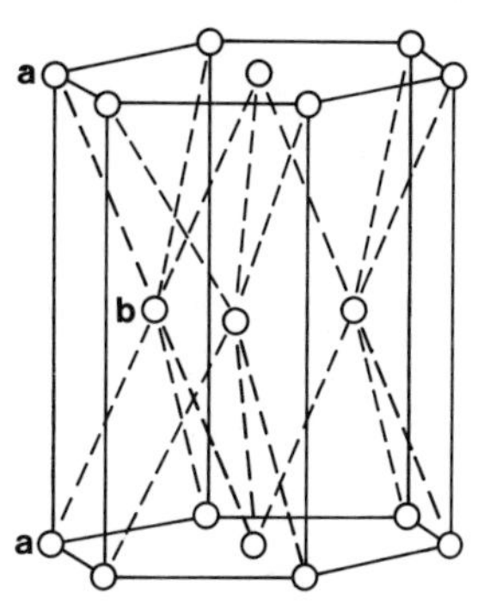

FIGURE 6.8. Conventional unit cell for HCP structure. Broken lines represent the nearest neighbors of the "b" layer in "a" layers.

Many compounds of the type AB, AB_2, A_2B, and so on have structures that can be considered as having the A or B atoms or ions in close-packed positions with the other atoms in the spaces in between them. As we see later, there are two different sorts of vacant spaces in the close-packed structures, referred to as *tetrahedral* and *octahedral* sites. We need to know how many and what sort of these there are in each close-packed structure.

Sites in Close-Packed Structures

A tetrahedral site (so named since the coordination number of the site is four) occurs every time a sphere sits on the depression on top of three others (see Figure 6.9).

An octahedral site occurs when there are four spheres touching in a plane to form a space, with one sphere above and one sphere below the space. Notice that the plane will always be at an angle to the plane of a close-packed layer, because no four spheres form a square in the plane of our close-packed layers. The octahedral site is shown in Figure 6.10.

To describe crystal structures it is useful to know the number of positions of each sort of site in the close-packed unit cells. Figure 6.11 shows the octahedral sites in the face-centered cubic (≡CCP) unit cell. One is at the center,

FIGURE 6.9. Two views of a tetrahedral site in close-packed layers.

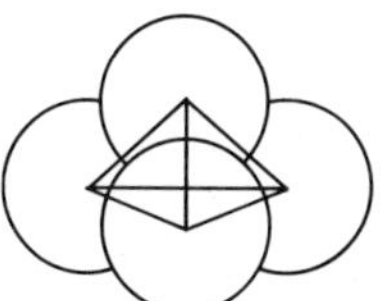

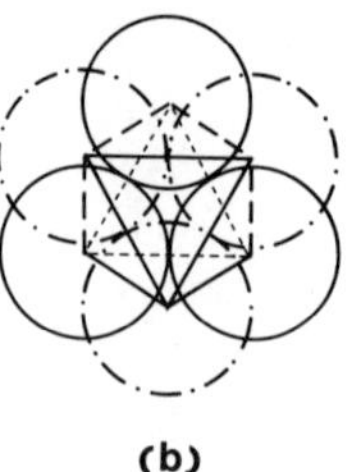

(a) (b)

FIGURE 6.10. Two views of an octahedral site.

its six nearest neighbors being the spheres at the center of the six faces. This one is in the unit cell. The others are at the midpoints of the edges, and for one of these, the nearest neighbors that are in adjacent unit cells are shown in Figure 6.11. There are twelve such octahedral sites, each shared between four unit cells (four cubes meet along a side), giving a total of

$$1 + 12\left(\frac{1}{4}\right) = 4$$

octahedral sites per unit cell.

Figure 6.12 shows one of the tetrahedral sites in the FCC lattice. If the whole unit cell is divided up equally into eight small cubes, the sides of which are half the size of the sides of the unit cell, it can be seen that there are spheres or lattice positions at alternate corners of the smaller cubes, forming a tetrahedral arrangement. The central point of these smaller cubes is a tetrahedral site, and there are eight of them per unit cell.

FIGURE 6.11. Octahedral holes in the FCC unit cell.

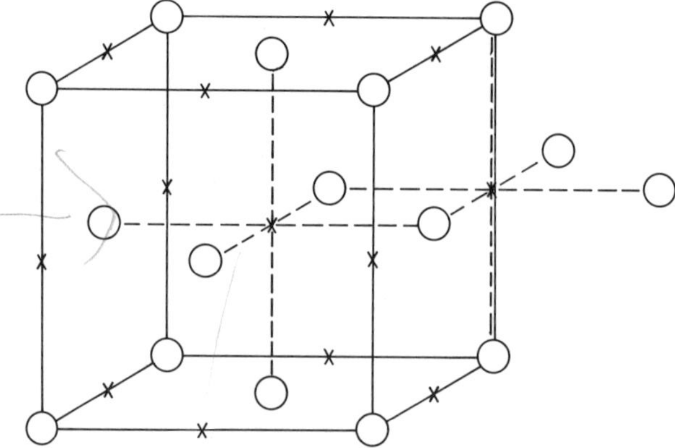

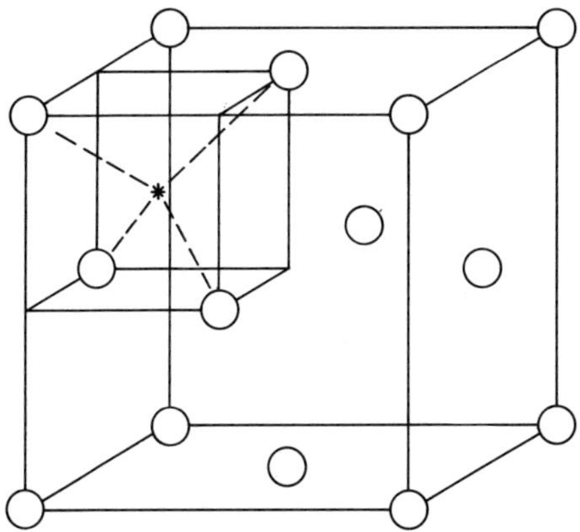

FIGURE 6.12. One of the eight tetrahedral sites in the FCC lattice.

Hence, to summarize, our conventional unit cell for the CCP or FCC structure has four lattice points per unit cell, four octahedral sites, and eight tetrahedral sites.

In a similar way, the octahedral and tetrahedral sites in the HCP lattice may be located and the number per unit cell worked out. Rather than do this we simply quote the result. There are six lattice positions per unit cell with six octahedral sites and 12 tetrahedral sites. Here the unit cell allows us to specify where the sites are relative to the close-packed layers. Figure 6.13 illustrates this and allows us to describe some structures.

FIGURE 6.13. Schematic of close-packed layers and position of sites in HCP lattice. Each line represents an equal number of the spheres or sites.

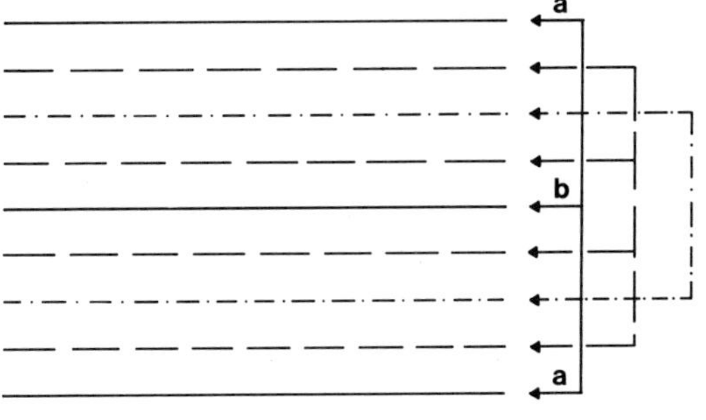

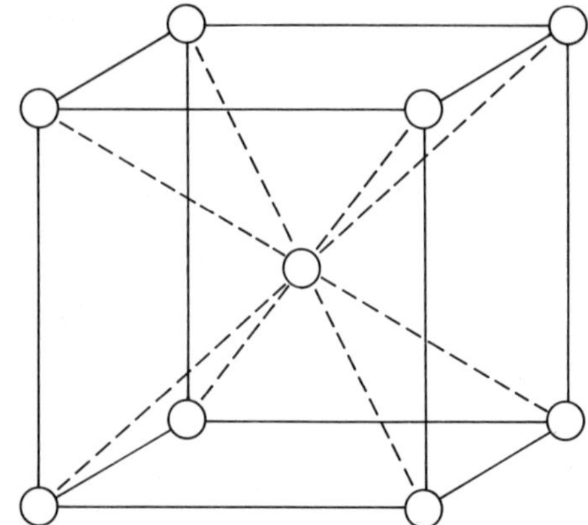

FIGURE 6.14.
Unit cell for body-centered cubic lattice.

Body-centered Cubic Lattice There is another simple common lattice type, called body-centered cubic (BBC) depicted in Figure 6.14, useful in describing solid-state structures. The body-centered cubic lattice uses less of the available space than the close-packed lattices. It uses only 68 percent, compared to 74 percent for the close-packed structures, that is, if the lattice points are spheres and are placed as close as possible to each other and still retain the body-centered cubic structure. The coordination number of each sphere in this case is eight. Notice that a sphere at a corner of Figure 6.14 will have as nearest neighbors the ones at the body-center of the eight unit cells that meet at that corner.

The Diamond Lattice The diamond structure has an atom at the face-centered cubic positions (not close-packed, but with some extra space in between) and the same atoms at half of the tetrahedral sites (every alternate one) (see Figure 6.15). It

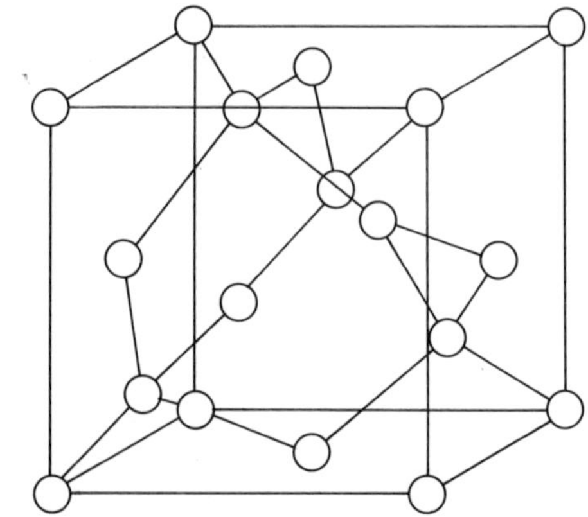

FIGURE 6.15.
The diamond lattice.

is not a close-packed structure since if it were, atoms of the same size would not fit in the tetrahedral holes.

Structures of Elements and Compounds

We have now covered enough ground to be able to describe a good many solid-state structures. However, it is well to remember that there are many other lattices than those we have described. In particular, there are many that have less symmetrical unit cells (i.e., distorted cubes), such that no two of the three mutually perpendicular edges have the same length or ones in which the angles between the edges are not right angles or both.

The Elements

Most of the elements are metallic, and most of these under normal conditions of temperature and pressure adopt one of four structures: (a) face-centered cubic, (b) hexagonal close-packed, (c) body-centered cubic, or (d) diamond lattice. The first two of these are closely packed structures and use all of the available space, while the latter two do not. We have seen the close-packed structures use 74 percent of the available space while the BCC uses 68 percent and the diamond lattice a lot less (only ca. 34 percent).

Figure 6.16 shows the structures most commonly (under normal conditions) adopted by the elements. Including the lathanides and the actinides, which are omitted from the table, well over half of the elements adopt close-packed structures in the solid state. Among those that do not are the alkali metals (all BCC) and some of the transition metals—mostly those with three or four d-electrons (again, nearly all BCC). The reasons for this are not very clear. It seems, however, that when there are few valence electrons the metal ions tend to have eight nearest neighbors (BCC), whereas when there are many, there are 12 nearest neighbors (closed-packed structures). One property of metals seems to correlate well with the number of valence electrons. Metals with a few valence electrons (s and p) outside closed shells tend to be rather soft. Those with some s- and p-electrons and also quite a number of d-electrons seems to be able to use the latter for forming additional directional bonds, that make the metals fairly hard.

The other elements in the periodic table, nonmetals and rare gases, tend to have a greater variety of structures. The rare gases, that is those that solidify, all form close-packed structures. There are in effect no valence electrons and the solids are held together by van der Waals forces.

FIGURE 6.16. Most common structure for the elements (the lanthanides and actinides have been omitted); M means molecular crystal and O, some structure other than the BCC, HCP, FCC, and D (diamond) structures.

H M						
Li BCC	Be HCP					
Na BCC	Mg HCP					
K BCC	Ca FCC	Sc HCP	Ti HCP	V BCC	Cr BCC	Mn O
Rb BCC	Sr FCC	Y HCP	Zr BCC	Nb BCC	Mo BCC	Tc HCP
Cs BCC	Ba BCC	La FCC	Hf HCP	Ta BCC	W BCC	Re HCP

*grey tin

Some of the nonmetals form molecular crystals, where the element is in the form of molecules, most often diatomic, in the solid. The molecules are then held together with the weak van der Waals forces. Some of the nonmetals, however, such as carbon and silicon, form lattices (i.e., the diamond lattice) where the atoms are bound together throughout the solid by strong covalent bonds.

Ionic Crystals

Many ionic crystal lattices can be described in terms of putting one sort of ion in the close-packed positions of either the cubic or the hexagonal close-packed lattices and the other ion in some or all of the tetrahedral or octahedral sites. Usually the cations have only anions as nearest neighbors and vice versa, although sometimes this is not so, giving the lattices different physical properties. The lattice is most strongly bound if the former situation prevails.

Other ionic crystal lattices do not take any of these structures, and we describe just a few of them here based on cubic lattices, which have a somewhat more open structure than the close-packed ones. Three of the most common simple ionic crystal lattices are shown in Figure 6.17. They are described in terms of placing the opposite sort of ions in the holes in the FCC close-packed structure (see Figure 6.17 caption). Table 6.2 lists the relevant crystal-structure data—number of cations and anions per unit

										He
					B O	C D	N M	O M	F	Ne FCC
					Al FCC	Si D	P M	S M	Cl M	Ar FCC
Fe BCC	Co HCP	Ni FCC	Cu FCC	Zn HCP	Ga O	Ge D	As M	Se M	Br M	Kr FCC
Ru HCP	Rh FCC	Pd FCC	Ag FCC	Cd HCP	In O	Sn* D	Sb O	Te M	I M	Xe FCC
Os HCP	Ir FCC	Pt FCC	Au FCC	Hg O	Tl HCP	Pb FCC	Bi O	Po O	At	Rn

cell, their coordination numbers, and the derived empirical formula. The latter, M_xA_y, is obtained by taking the inverse ratio of the corresponding coordination numbers. For example, for the fluorite structure, since the coordination number of the cation is 8, one-eighth of each cation is associated with an anion and since the coordination number of the anion is 4, one-fourth of each anion is associated with a cation; hence the empirical formula is $M_{1/8}A_{1/4}$ or MA_2. Thus in the formula M_xA_y:

$$\frac{x}{y} = \frac{\text{coordination no. anion}}{\text{coordination no. cation}}.$$

Figure 6.18 shows the wurtzite structure (another form of ZnS). It is a structure obtained by putting cations in half of the tetrahedral holes of an HCP structure of anions.

Table 6.2 Crystal-structure data for structures derived from FCC close-packed structures

Crystal structure	No. of cations per unit cell	No. of anions per unit cell	Coordination no. of cation	Coordination no. of anion	Empirical formula
Rocksalt (NaCl)	4	4	6	6	MA
Zinc blende (ZnS)	4	4	4	4	MA
Fluorite (CaF_2)	4	8	8	4	MA_2

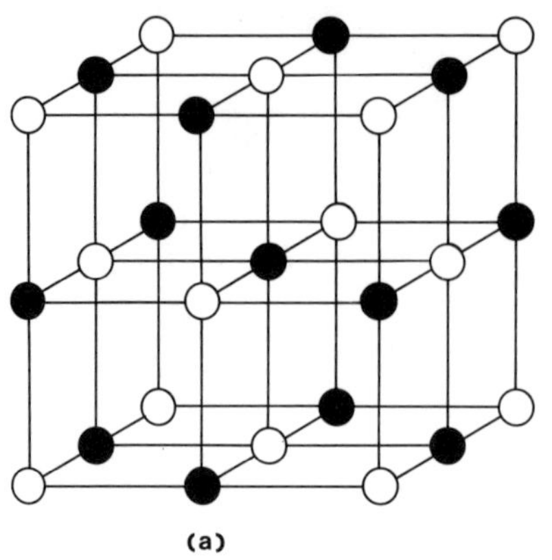

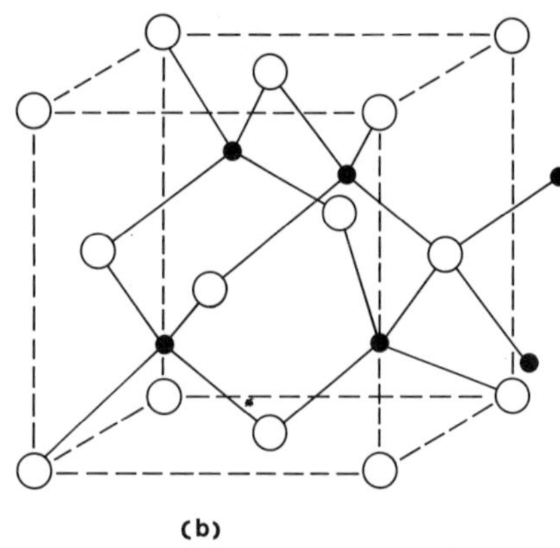

FIGURE 6.17.
Crystal structures derived from FCC structure. (a) Rock salt (NaCl) anions in FCC positions, cations in all octahedral holes; (b) zinc blende (ZnS) anions in FCC positions, cations in half the tetrahedral holes; (c) fluorite (CaF_2) cations in FCC positions, anions in all tetrahedral holes.

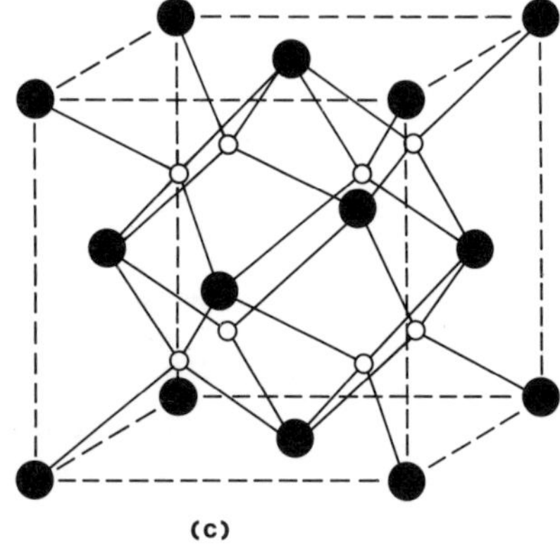

Thus far we have only shown the positions of the layers of the CP spheres and sites for HCP structures (see Figure 6.13). The layers occupied in the wurtzite structure are shown in Figure 6.19. Since this structure is one form of zinc sulfide, we have shown the layers in the other form (zinc blende) beside it in Figure 6.19. The difference

FIGURE 6.18.
Crystal structure (wurtzite) obtained by putting cations in half the tetrahedral sites of an HCP structure of anions.

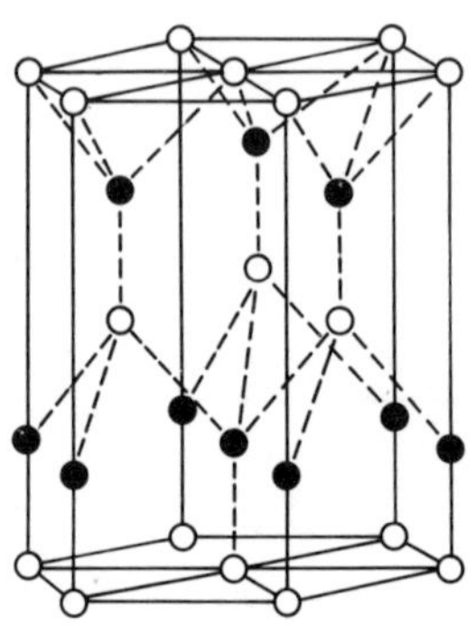

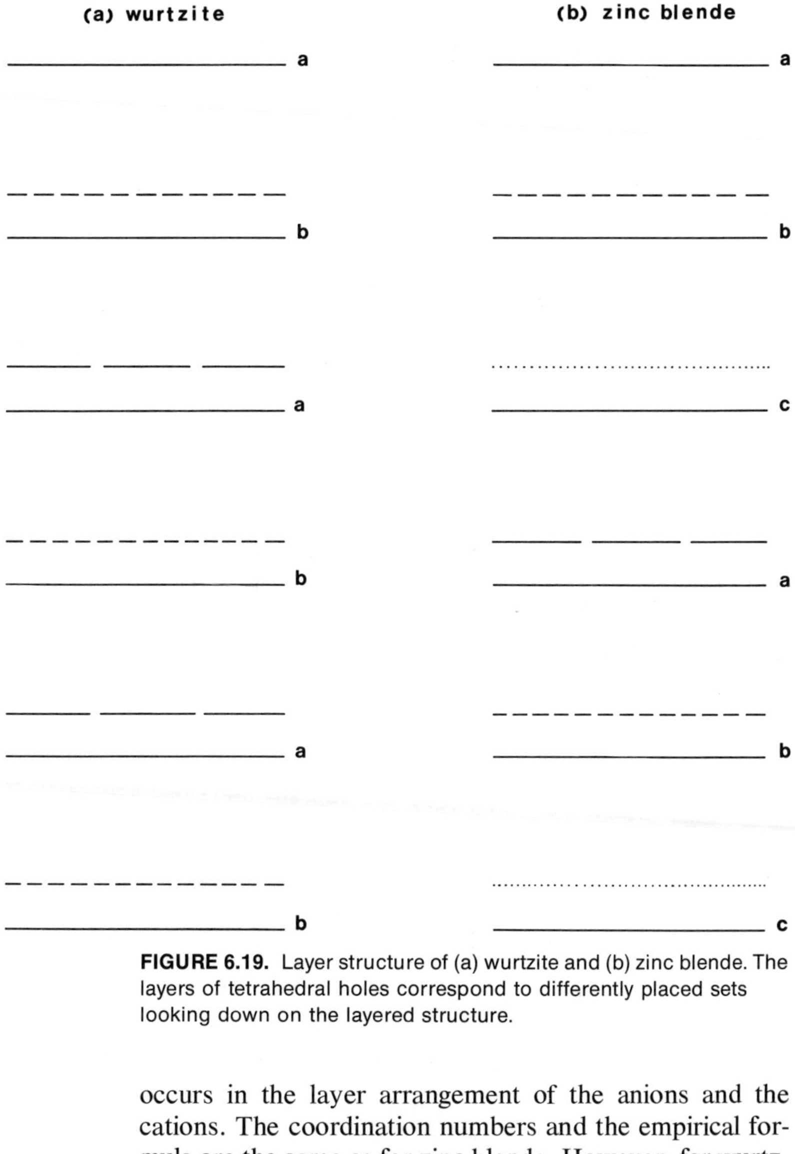

(a) wurtzite

a
b
a
b
a
b

(b) zinc blende

a
b
c
a
b
c

FIGURE 6.19. Layer structure of (a) wurtzite and (b) zinc blende. The layers of tetrahedral holes correspond to differently placed sets looking down on the layered structure.

occurs in the layer arrangement of the anions and the cations. The coordination numbers and the empirical formula are the same as for zinc blende. However, for wurtzite there are six of each ion in the unit cell. It is interesting to note that these structures use the same amount of space in the hard-sphere description and the coordination numbers are the same, but, as we have seen, the structures are different.

Figure 6.20 shows a simple cubic lattice structure. Here the anions in the unit cell are at the corners of a cube with the cations in all positions having a coordination number of 8 with respect to the anions. The structure can be

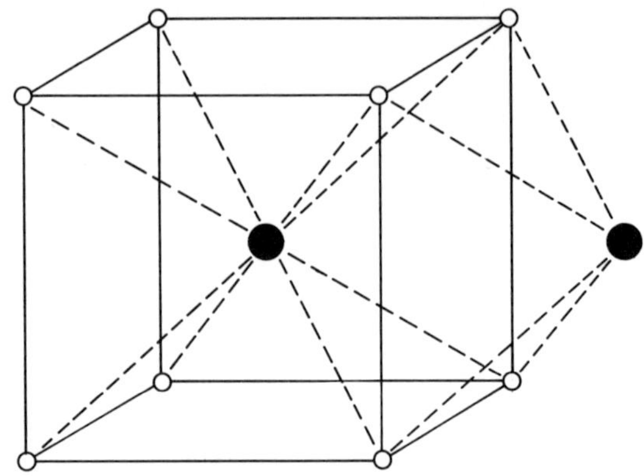

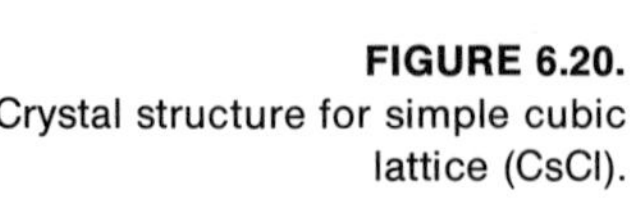

FIGURE 6.20.
Crystal structure for simple cubic lattice (CsCl).

thought of as two simple cubic lattices, one placed with lattice points as far away as possible from the lattice points of the other. It is often incorrectly termed the *body-centered cubic lattice*. However, "lattice" infers that each unit is the same and in a common example, cesium chloride, we have chloride ions at the corners of the cube and a cesium ion at the body center. The crystal structure is correctly described by putting a Cs^+Cl^- unit at the lattice points of a simple cubic lattice with all the Cs^+Cl^- units pointing in the same direction—the one shown in Figure 6.20. For this structure the coordination number of both ions is 8, and thus the empirical formula is MA.

Radius Ratio and Crystal Structures

There is a general correlation between the crystal structure a particular ionic compound has and the relative radii of the cation and anion. The octahedral and tetrahedral holes in close-packed structures have different sizes (the former being larger than the latter), and the structures tend to be those for which there is a snug fit. To overemphasize this we could say that a small cation would "rattle around in a large hole" and at the other extreme a large cation would tend to force the close-packed structure of anions too far apart in a small hole to get a rigid structure. Outside the two sites mentioned above we have cubic holes in the simple cubic lattice for very large cations and triangular ones for very small cations.

It is relatively simple to calculate the optimum size of a cation that will fit into the various coordination sites. For example, an octahedral site has its diameter determined by four spheres touching as illustrated in Figure 6.21. Simple geometry then leads us to

$$2(r_c + r_a)^2 = (2r_a)^2 \qquad (6.1)$$

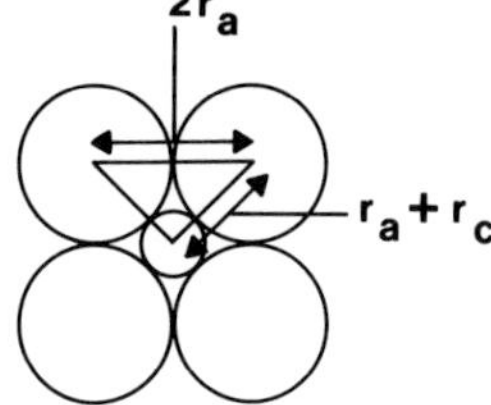

FIGURE 6.21.
Inner circle represents the octahedral hole in an octahedral site. In the structure there is another anion directly above and below the cation.

where r_c is the radius of the cation and r_a is the radius of the anion. Solution of this equation gives

$$\frac{r_c}{r_a} = 0.414.$$

Similar calculations for sites of other coordination numbers lead to

$$\begin{aligned}\frac{r_c}{r_a} &= 0.732 \text{ for cubic coordination}\\ &= 0.225 \text{ for tetrahedral coordination.}\end{aligned}$$

The change in coordination number of a site for a particular cation seems to occur when the cation gets smaller than the "snug-fit" value, so that we may represent the changes in coordination number for cations in different crystal structures by the diagram that follows. The numbers represent r_c/r_a values at which the coordination number tends to change:

$$\text{Cubic (8)} \overset{0.732}{\rightleftharpoons} \text{octahedral (6)} \overset{0.414}{\rightleftharpoons} \text{tetrahedral (4)} \overset{0.225}{\rightleftharpoons} \text{triangular (3).}$$

Many crystal structures may be rationalized using this. Table 6.3 lists a number of common halides, oxides, and sulfides, together with the corresponding radius ratio and the crystal structure adopted. It should be noted that there are some "wild" deviations from the above rules, for example, CsF, but the general trends are followed.

There is no such thing as a perfect ionic crystal, which would involve complete transfer of the electron from the cation to the anion and no distortion of the spherical symmetry of the ions. In fact, there is a whole range of ionic crystals varying from nearly perfectly ionic to nearly molecular. An ionic crystal such as Na^+Cl^-, where the electronegativities of the respective ions are very low and very high, is close to being ionic (see Chapter 5). Others such as BeO, in which the Be occupy tetrahedral sites in the HCP lattice of the oxygens (see Figure 6.18) one BeO

Table 6.3 Crystal structures and radius ratios for some common ionic solids

Compound	Radius ratio	Coordination no. of cation	Structure
BeS	0.17	4	Zinc blende
BeO	0.22	4	Wurtzite
LiF	0.44	6	NaCl
NaF	0.70	6	NaCl
KF	0.98	6	NaCl
RbF	1.09	6	NaCl
CsF	1.24	6	NaCl
LiCl	0.33	6	NaCl
NaCl	0.52	6	NaCl
KCl	0.73	6	NaCl
RbCl	0.82	6	NaCl
LiBr	0.31	6	NaCl
NaBr	0.49	6	NaCl
KBr	0.68	6	NaCl
RbBr	0.76	6	NaCl
LiI	0.28	6	NaCl
NaI	0.44	6	NaCl
KI	0.62	6	NaCl
RbI	0.69	6	NaCl
MgO	0.46	6	NaCl
CaO	0.71	6	NaCl
SrO	0.81	6	NaCl
BaO	0.96	6	NaCl
MgS	0.35	6	NaCl
CaS	0.54	6	NaCl
SrS	0.61	6	NaCl
BaS	0.73	6	NaCl
CsCl	0.93	8	CsCl
CsBr	0.87	8	CsCl
CsI	0.78	8	CsCl

distance is always shorter than the other three, and so the solid tends to have molecular units in the crystal. Cases such as BeO are intermediate between nearly pure ionic and covalent. At the extreme, iodine crystals consist of molecular units bound by weak van der Waals forces. Besides the evidence from the unequal bond lengths, the properties of the different solids, tabulated in Table 6.1, indicate the variations from nearly ionic to nearly covalent in the structure of ionic solids.

Covalent Solids

Under the heading of covalent solids we include all solid-state structures except metals, molecular lattices, and ionic solids. The latter are different from many covalent solids in that they tend to have very high melting points. However, this is not a very valid basis for distinction, since some ionic solids have lower melting points (generally as the tendency toward molecular units increases), and some covalent solids have extremely high melting points (if they are bonded completely by covalent bonds; e.g., as in diamond structures). The ion in an ionic crystal may also be a molecular ion, within which covalent bonds are holding the constituent atoms of the ion together. Thus perhaps a better distinguishing feature between the two classes of solids would be that ionic crystals are those in which the electrostatic attraction of differently charged ions is the main obstacle to be overcome to change the solid state into the liquid; when this is not so, we have covalent solids or molecular lattices.

There are three general classifications of covalent structures: (a) three-dimensional giant molecules, (b) two-dimensional molecules or layer lattices, and (c) one-dimensional molecules or chain lattices. We now consider briefly the structure and properties of each.

Three-dimensional Giant Molecules

We have already illustrated the diamond lattice in Figure 6.15, which as you may recall consists of a face-centered cubic lattice with units of equal size at half the tetrahedral holes. In diamond each lattice point is a carbon atom, with each covalently bound to four neighbors. The coordination number is four for every carbon, which is tetrahedrally surrounded by four others in the lattice. The characteristic properties of such a lattice are: (a) its hardness—to fracture diamond a large number of covalent bonds have to be broken in any given direction—and (b) its nonconduction of electricity—the valence electrons are all tied up in directional covalent bonds. Notice, however, that there is more empty space in the diamond lattice than the close-packed structures.

Two other common examples of three-dimensional giant molecules are carborundum and silica. Carborundum is silicon carbide and has the zinc-blende structure with silicon atoms in place of zinc and carbon atoms in place of sulfur, or vice versa. The atoms in the solid are again bound throughout by covalent bonds giving carborundum its characteristic property of hardness. Silica, one of the many forms of silicon dioxide, has silicon atoms at both the zinc and sulfur positions in the zinc-blende structure and oxygen atoms half way in between the silicon atoms.

Thus the coordination number of oxygen is 2 and that of silicon, 4, corresponding to the empirical formula SiO_2. The structure may be described as an infinite three-dimensional lattice of SiO_4 tetrahedra with every oxygen atom shared by two silicon atoms. Again the covalent bonds in three dimensions give rise to the property of hardness.

Layer Lattices

Layer lattices are characterized by two-dimensional arrays of covalently bonded units (planes or layers), with these layers in turn loosely bound to each other by the weaker van der Waals forces or weak ionic attractions. They tend to form flaky solids because the layers can slide over one another fairly easily; hence many of them are used as lubricants. Graphite, another form of solid carbon, is a typical example. It consists of infinite layers of carbons bound together by covalent bonds into hexagons shown as follows:

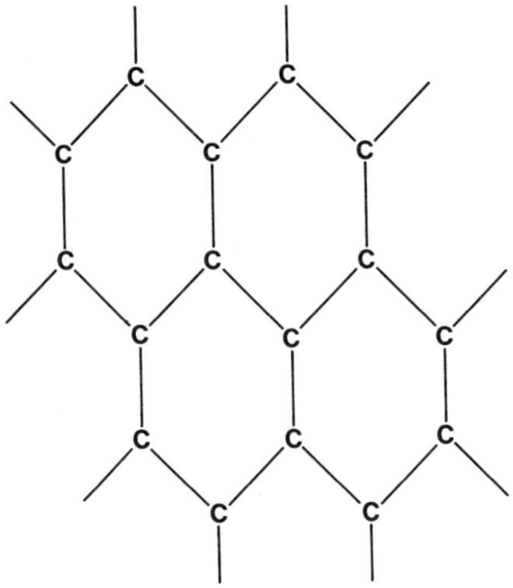

There is then one valence electron ($2p_z$ if the z-axis is perpendicular to the layer) on each carbon, and these bind the layers together by weaker forces than the covalent bonds within the layers. The electrons binding the layers together are reasonably mobile. This is shown by the fact that the electric conduction perpendicular to the layers is about 10^{-4} times as great as that parallel to the layer; thus the conduction is very anisotropic. The structure of graphite is shown in Figure 6.22.

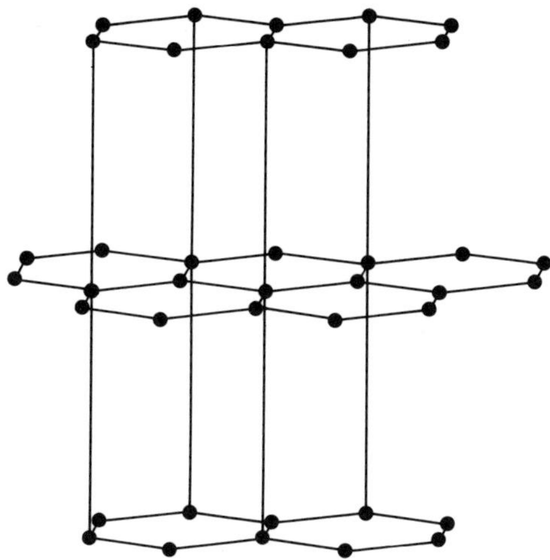

FIGURE 6.22. The structure of graphite.

Another layer lattice is formed by another form of silica. Here SiO_4 tetrahedra are bound to each other in a plane by sharing only three of the four oxygens between tetrahedra. The resulting layers have excess negative charges, which are bound together by positively charged cations. Talcum powder has silica layers like this bound together by Mg^{2+} ions. Micas have similar structures with some of the silicon atoms replaced by aluminium and with K^+ and Ca^{2+} ions binding the layers together.

Those ionic crystals that tend to have a lot of covalent bonding also form layer-like lattice structures. For example, cadmium iodide has the structure in which the iodide ions form an hexagonal close-packed lattice with the cadmium ions in half of the octahedral sites. This half, that is three in the conventional unit cell (see Figure 6.13) occupy all of the octahedral sites between every alternate pair of layers of iodide ions leaving the ones in between vacant (see Figure 6.23). The bonding between the cadmium and iodine atoms has a large amount of covalent character, forming tightly bound layers. There is a slight negative charge on the top and bottom of each layer, which binds them very weakly and thus again we get a flaky layer like crystal. Notice that the coordination number of cadmium is six in this structure (they are at octahedral sites) but the coordination number of iodine is only three (the cadmiums occur only on one side of iodine) giving the empirical formula CdI_2.

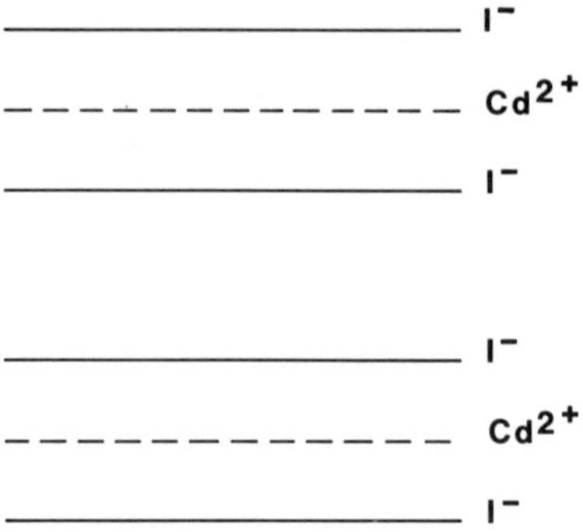

I^-
Cd^{2+}
I^-

I^-
Cd^{2+}
I^-

FIGURE 6.23. Layer structure of cadmium iodide.

Chromic chloride and aluminum hydroxide have solid state structures slightly different to that of cadmium iodide. The anions again form HCP structures but the cations fill only two thirds of the octahedral sites between every alternate pair of layers. Thus the coordination number of the cation is again 6, but that of the anions is 2 ($\frac{2}{3}$ of 3), giving the formulas $CrCl_3$ and $Al(OH)_3$. The aluminum hydroxide structure is shown in Figure 6.24.

Chain Lattices

These are structures in which the most strongly bound atoms form endless chains or strands, which in turn are held together, often in a twisted, crazy, mixed up mess, by much weaker van der Waals or electrostatic forces. This results in the solids with chain lattices having fibrous properties. There is a great range of naturally occurring substances, even some solids that might be classed as

FIGURE 6.24. Structure of aluminum hydroxide.

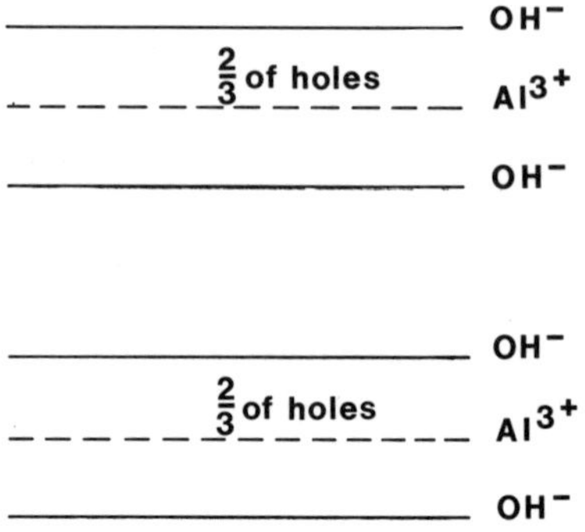

OH^-
$\frac{2}{3}$ of holes Al^{3+}
OH^-

OH^-
$\frac{2}{3}$ of holes Al^{3+}
OH^-

ionic crystals and also very many synthetic polymers. For the latter, the fibrous property is often lost by the method of molding, but the basic unit is still often a long chain structure.

Examples of some naturally occurring chain lattices are plastic sulfur and cellulose. Plastic sulfur consists of infinite chains of

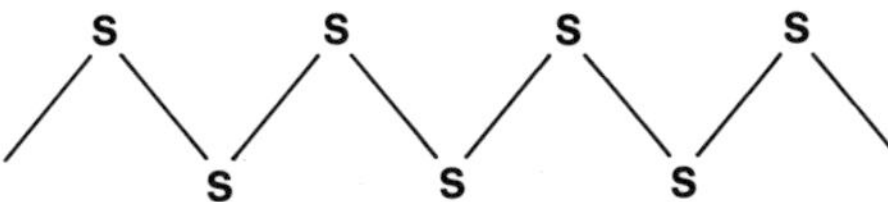

bound together by weak van der Waals forces. Cellulose consists of infinite chains of sugar molecules joined together. Each is joined to two others by the elimination of water, forming covalently bonded chains. The sugar molecule is:

and the chains may be written as:

Copper II chloride is an example of a chain lattice. It has long chains of

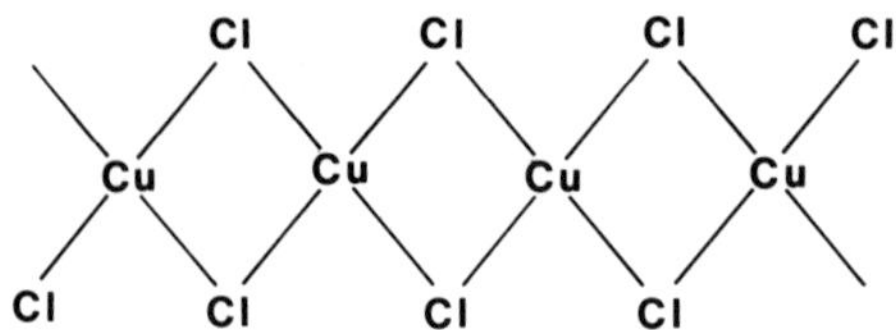

with two chlorine atoms in neighboring chains close enough to each copper atom to yield a coordination number of 6. The chlorine atoms are around each copper atom in the shape of a distorted octahedron. Even within each chain two chlorine atoms are closer than the other two to each copper atom.

Yet another class of silicates are those in which SiO_4 tetrahedra share only two of the oxygen atoms with adjacent tetrahedra. Some of these have single strands of tetrahedra and some have double strands. The charge on the repeating unit of these structures is easy to calculate since every oxygen atom not shared by two tetrahedra needs an extra electron to complete its octet. Thus when two are shared the structural formula is:

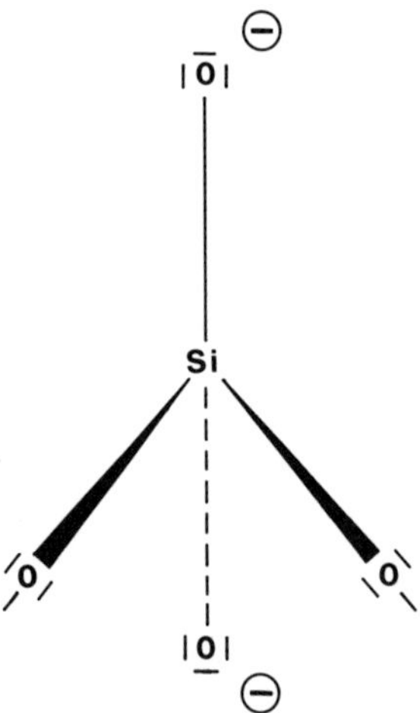

If we represent this by:

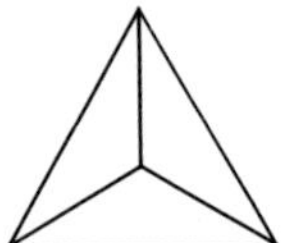

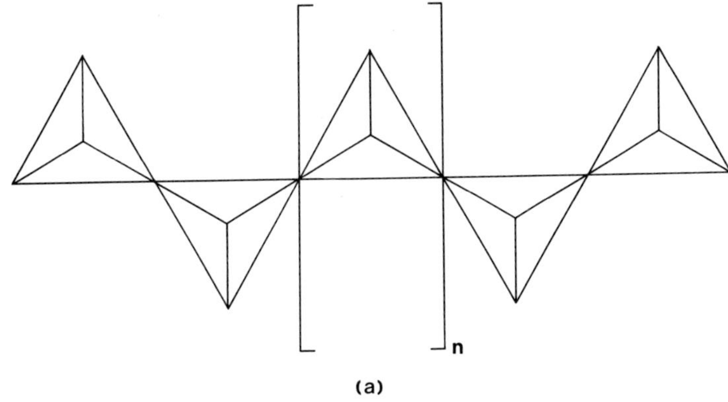

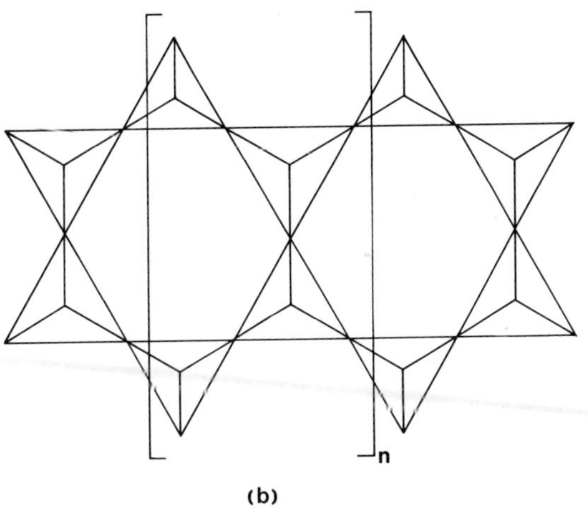

FIGURE 6.25. Silicate chain. (a) Single-stranded structure; (b) double-stranded structure.

we can write the two structures shown in Figure 6.25, with the repeat unit in brackets. It is then easy to work out that the formula for the repeat unit is

SiO_3^{2-} for single-stranded
and
$Si_4O_{11}^{6-}$ for double-stranded structures.

These formulas of course neglect end effects in the very long chains. An example of a chain silicate is asbestos. It has double-stranded chains loosely joined together by Na^+, Fe^{2+}, Fe^{3+} and OH^-. The empirical formula for one of the common forms is $Na_2Fe_5(OH)_2Si_8O_{22}$.

Many synthetic polymers have a basic unit that is polymerized into a long chain and the polymer then fabricated into its final form. The simplest one of these is polythene, in which ethylene is polymerized to give long chains as

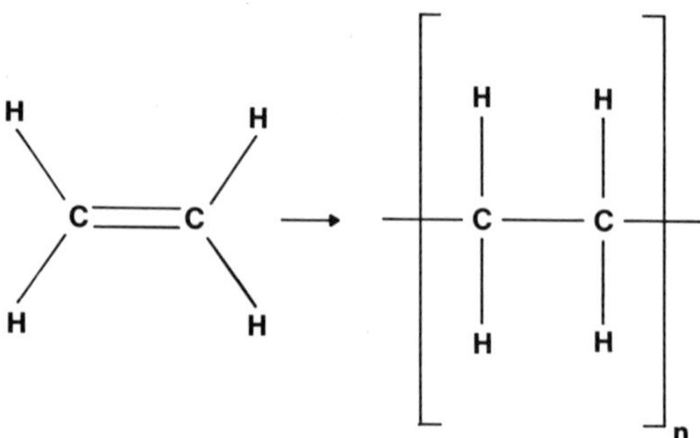

The chains are then fabricated into the desired shape by some suitable method. Generally polymers do not possess the fibrous property of other loosely bound chain lattices. Table 6.4 lists the monomers (e.g., ethylene) that polymerize to give the appropriate polymer (e.g., polythene), given

Table 6.4 Some common polymers and their source monomers

Monomer	Polymer	Some uses
$CH_2{=}CH_2$	Polythene	Films, containers
$CH_2{=}CH(C_6H_5)$	Styrofoam	Insulating, molding
$CH_2{=}CHCl$	PVC (polyvinyl chloride)	Tubing
$CH_2{=}CCl_2$	Gladwrap	Films, wrapping
$CF_2{=}CF_2$	Teflon	Seals, bearings, laboratory ware
$CF_2{=}CClF$	Kel F	Lubricants
$CH_2{=}CHCN$	Orlon	Wool substitute
$CH_2{=}CCH_3COOCH_3$	Perspex	Glass substitute, molding
$CH_2{=}CClCH{=}CH_2$	Neoprene	Synthetic rubber
Glycol + a dicarboxylic acid (a polyester)	Terylene	Clothing
Hexamethylene diamine + adipic acid (a polyamide)	Nylon	Clothing
A di-isocyanate + a diol (a polyurethane)	Durethan	Foam
Phenol + formaldehyde	Bakelite	Household objects
Formaldehyde	Delrin	Bearings
$SiO_2(CH_3)_2$	Silicone rubber	Seals

as its common trade name. In making polyesters, polyamides, and polyurethanes, a variety of acids, alcohols, and other reagents have been used, and these are not specified exactly in the table.

Molecular Lattices

The last type of lattice we wish to discuss is that in which the lattice unit is small, most often a normal molecule. These are bound together by weak van der Waals forces, dipole–dipole forces, or sometimes by somewhat stronger hydrogen bonds. They are characterized by not being very hard and having fairly low melting points.

Typical examples of molecular lattices held together by van der Waals forces are the solid forms of the halogens, chlorine, bromine, and iodine. The structures of these are quite similar and show some interesting features. For example, in solid chlorine every chlorine atom has *one* (and only one) neighbor at a distance of 202 pm away. In the same plane there are others 334 pm away, and in the layers above and below a particular plane the atoms are 369 pm and 373 pm away. The first distance (202 pm) is close to the gas-phase diatomic molecule internuclear distance. The value 334 pm suggests that there is a small amount of interhalogen molecule bonding in the layer and the values 369 pm and 373 pm between layers suggest very little bonding between layers. Thus the crystal consists of molecular units bound together by weak van der Waals forces with a small amount of interhalogen molecule bonding.

Many other molecular lattices are similar to the halogens in that the molecules are stacked in a regular fashion in the crystal with the molecular unit bound essentially as it is in the gas phase and neighbors further than normal covalent distances away. For example, planar aromatic molecules have a structure in which the molecules are stacked in planes not parallel to each other.

In other molecular crystals the molecular unit is often not the one we expect. For example one form of sulfur and one of phosphorus have the molecular units:

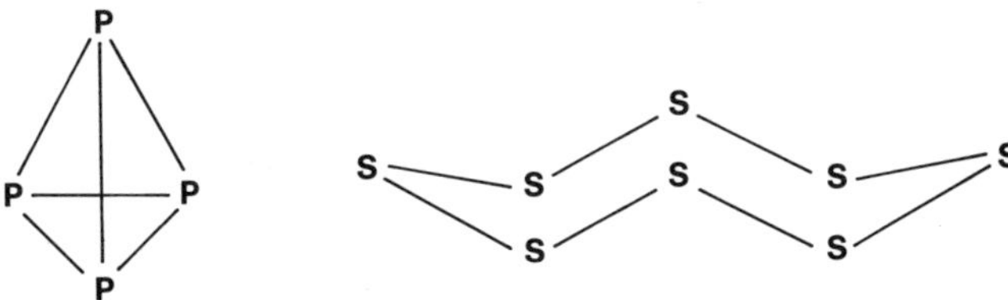

while the phosphorus oxides P_2O_3 and P_2O_5 have structures:

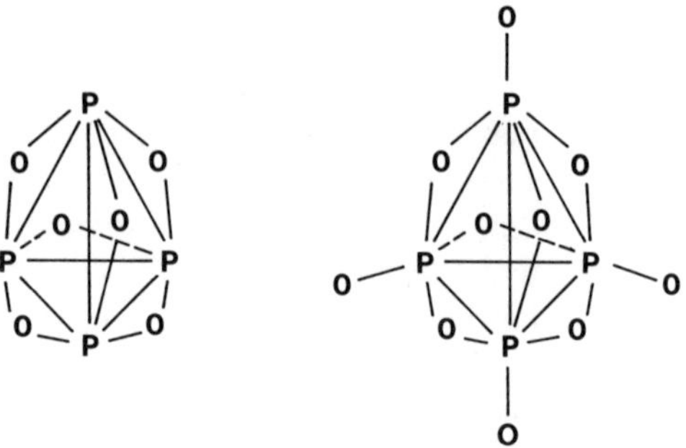

as the molecular units.

In other molecular lattices, in addition to van der Waals forces and dipole–dipole attractions, hydrogen bonds play a very large part in determining the crystal structure. Many of the molecules where this is so are biologically very important. One of them, water, is probably the most abundant molecule on earth.

Ice has a structure with the oxygen atoms in the occupied sites in the wurtzite structures. That is, oxygen atoms are at both the Zn and S positions so that each oxygen is tetrahedrally surrounded by four others. The hydrogens

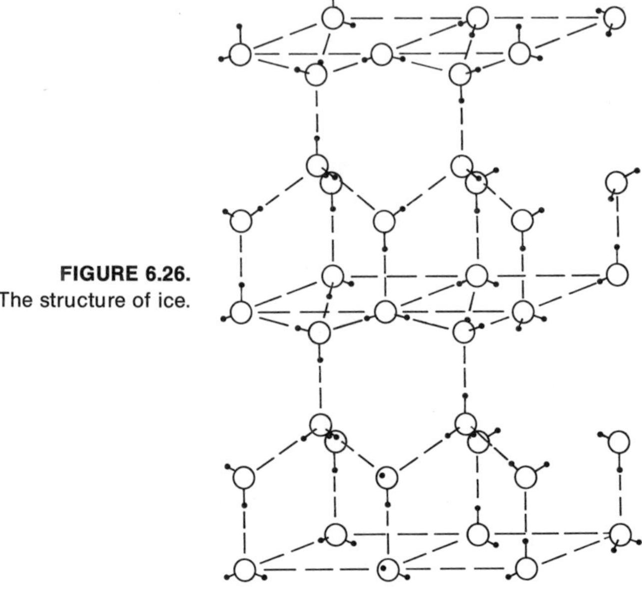

FIGURE 6.26. The structure of ice.

are along the O—O directions with two closer to each O than the other two. The special characteristics of this structure are that it is very open and that the hydrogens quite frequently change the oxygen to which they are closest. Two are covalently bonded to an oxygen, with two from two neighboring water molecules being hydrogen-bonded. Because of the hydrogen bonds, ice is held together quite strongly. Also because of the open structure, the density of ice is less than that of water. When ice melts, not all of the hydrogen bonds are broken, but the rigidity of the solid is lost and the structure is not so open, thus having a higher density. The structure of ice is shown in Figure 6.26.

Many other solids have hydrogen bonds holding them together more tightly than would otherwise be. Some organic acids form hydrogen bonded dimers in the gas phase; in other words:

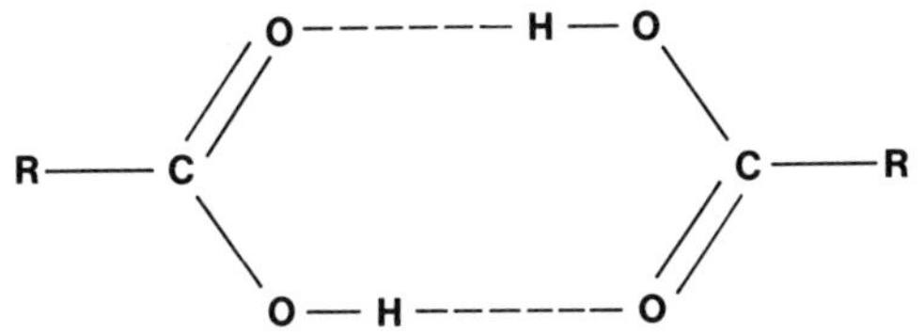

but in their solids generally form long hydrogen bonded chains as:

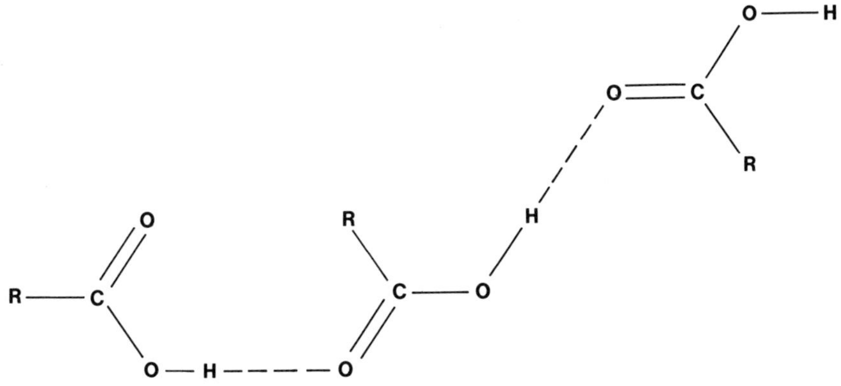

A Biological Macromolecule, DNA

We wind up the survey of solid-state structures by describing briefly one biological macromolecule. Hydrogen bonds play a major part in determining the conformation of this molecule. Deoxyribonucleic acid (DNA) consists of two long-chain molecules twisted together in the form of a double helix. The long chains consist of sugar molecules (five membered rings) joined together by phosphate groups. On each sugar entity in place of one hydrogen is a purine or pyrimidine base. The double-stranded helix structure of the two long chains is formed by a base from

FIGURE 6.27. Hydrogen-bonded purine and pyrimidine base pairs found in DNA.

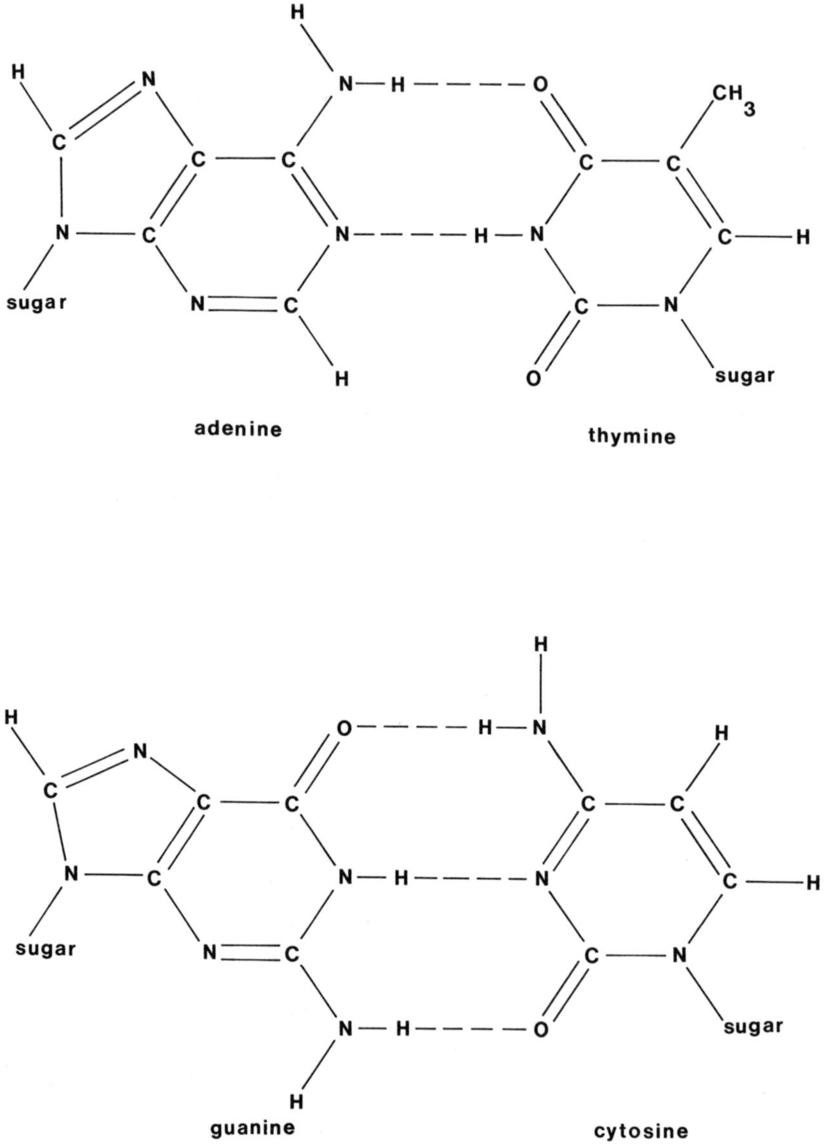

adenine thymine

guanine cytosine

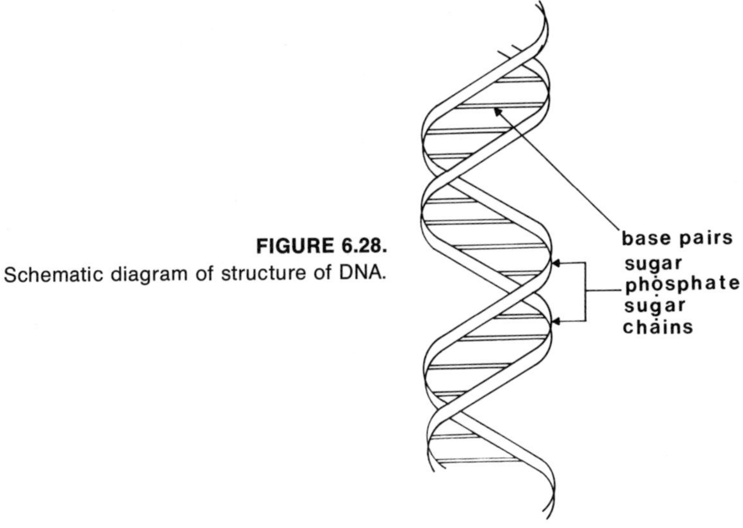

FIGURE 6.28.
Schematic diagram of structure of DNA.

one chain hydrogen bonding to a base from the other. The pairs of bases that will form these hydrogen bonds are very specific. It is always a pyrimidine base hydrogen bonded with a purine base, and vice versa. In fact, with the four common bases, adenine, thymine, guanine, and cytosine, it is the combinations shown in Figure 6.27 that are found.

A schmatic diagram of the molecule is shown in Figure 6.28.

Problems

6.1 Classify the following molecules as to their type of solid-state bonding:

a. acetylene
b. gold
c. hydrogen fluoride
d. carborundum
e. $CuCl_2$
f. polypropylene
g. CsBr
h. naphthalene

6.2 What structural feature is common to both hexagonal close-packed and face-centered cubic lattices?

6.3 Draw the conventional unit cell for the crystal structures of the following:

a. M^+X^- where the anions form a cubic close-packed lattice and the coordination number of the cations is 4
b. M^+X^- where the anions form a cubic close-packed lattice and the coordination number of the cations is 6

6.4 Draw a conventional unit cell of Na^+Cl^- showing clearly the location of the Na^+ ions and the Cl^- ions. How many cations are there in the unit cell, and what is the coordination number of these cations?

6.5 The structure of a compound of sodium, niobium and oxygen is as follows:

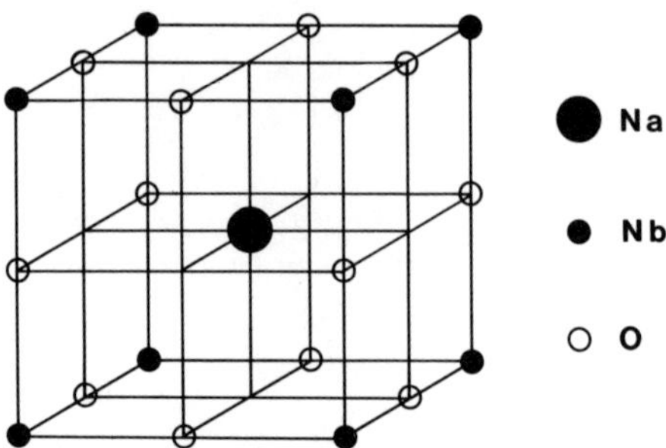

How many atoms of Na, Nb, and O are in the unit cell, and what are the coordination numbers of these atoms?

7 Experimental Methods of Valency

There are numerous experimental techniques used not only to identify and determine the geometry of molecules to varying degrees of accuracy, but also in some cases to provide values of such quantities as dissociation energies. These methods can be divided into two distinct types: (a) *spectroscopic* and (b) *diffraction* methods. Sometimes in the former method diffraction gratings are used to disperse the wavelengths of the radiation used, but the essential difference is that in spectroscopic methods we are studying the absorption or emission of radiation by atoms or molecules (in any phase), whereas in the diffraction method we are using the atoms and molecules (again in any phase) as the diffracting medium. Mass spectroscopy is another method for studying molecular structure, utilizing the separation of atomic or molecular ions of different mass-to-charge ratios in a magnetic field.

Spectroscopic Methods

In Chapter 3 we dealt at some length with the properties of electromagnetic radiation. Spectroscopic methods involve the absorption or emission of electromagnetic radiation. When this occurs the molecule undergoes a transition to a new energy level, the frequency of the radiation and the energy level difference being related by the Bohr frequency condition.

$$\Delta E = E_2 - E_1 = h\nu. \qquad (7.1)$$

Notice that this equation is for one photon so that the energy difference is for a single molecule.

To a good approximation the energy of a molecule may be written as

$$E = E_{\text{trans}} + E_{\text{rot}} + E_{\text{vib}} + E_{\text{elec}}. \tag{7.2}$$

where

E_{trans} is the energy of translation;
E_{rot} is the energy of rotation;
E_{vib} is the energy of vibration;
and E_{elec} is the electronic energy.

Equation 7.2 means that these different kinds of energy may be separated from each other, so that the total energy is a summation of them. Hence we may talk about rotational, vibrational, and electronic energy levels. As far as molecular spectroscopy is concerned we are not particularly interested in translational energy levels because the separation between them is very small and would yield no interesting molecular parameters. The separations between electronic energy levels are of the order of 10,000–50,000 cm^{-1}, those for vibrational energy levels, 200 cm^{-1} to 4,000 cm^{-1} and those for rotational energy levels, about 1 cm^{-1}. Thus we may build the vibrational energy levels on the electronic levels and then the rotational energy levels on the vibrational levels as shown in Figure 7.1. Because every molecule possesses what is called zero-point vibrational energy, even at zero degrees Kelvin, molecules never actually occupy the electronic levels marked E_1 and E_2 but are at least in the level marked $v = 0$ and $J = 0$.

Just as for the electronic energy levels in atoms, the molecular-energy levels are all described by quantum numbers. These are too messy to worry about for the electronic levels, but each vibrational level in each electronic state is described by a quantum number v that takes the values $v = 0, 1, 2, 3, \ldots$, and each rotational level in each vibrational state is described by a quantum number J that takes the values $J = 0, 1, 2, \ldots$. From Figure 7.1 we can see what changes in energy levels are involved for different energy regions. Not all molecules are in the lowest energy state at room temperature. In general, most molecules in a gas are distributed amongst the rotational levels of the lowest vibrational level, (higher vibrational levels are populated to a small degree) of the ground electronic state. Hence there are groups of transitions that

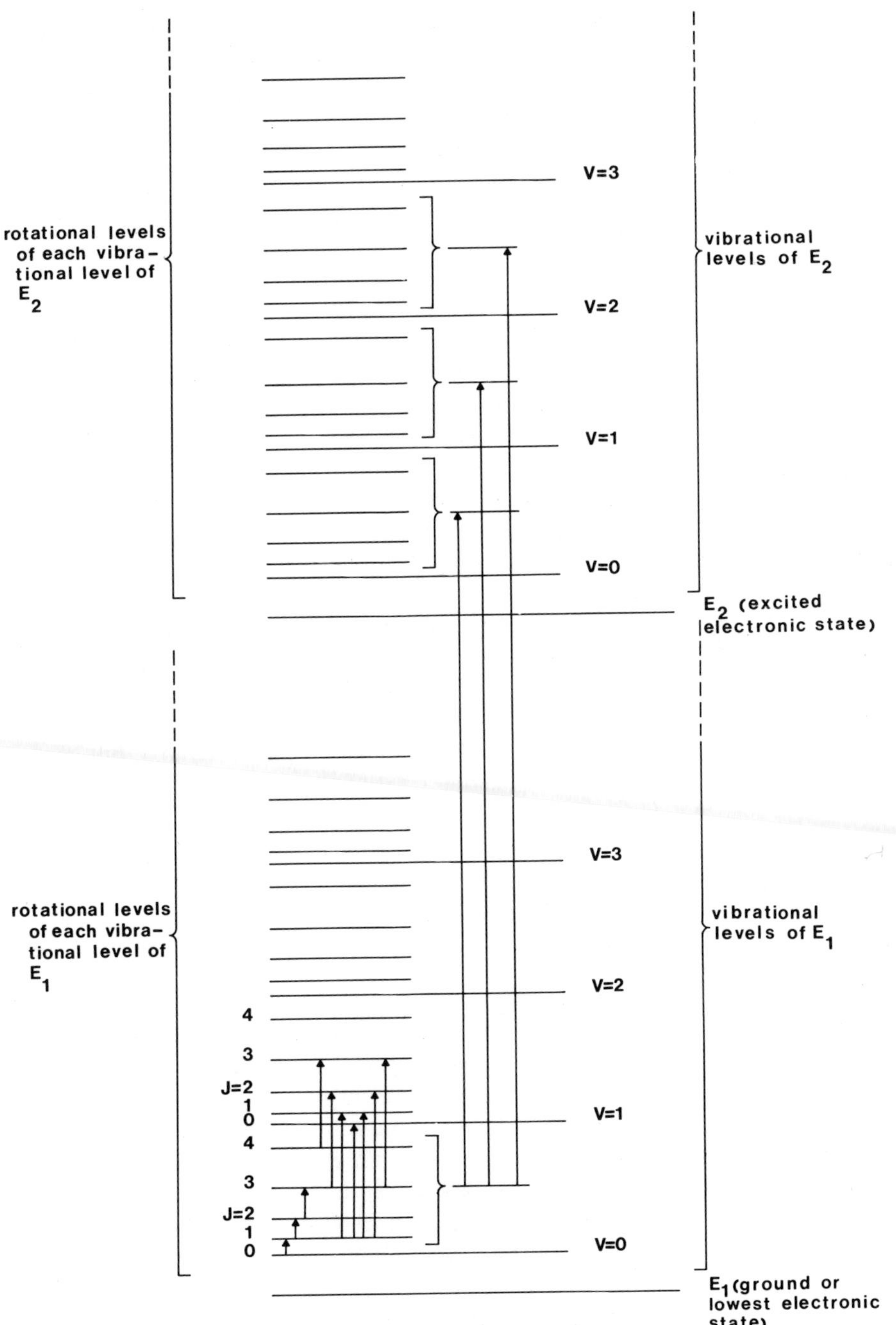

FIGURE 7.1. Energy levels of a molecule showing rotational, vibrational–rotational, and electronic–vibrational–rotational transitions.

can be observed in three different energy regions corresponding to three different regions of the electromagnetic spectrum. These are described in the text that follows.

Rotational Transitions

These occur between different rotational levels of the lowest vibrational level of the ground electronic state and are located between about 3mm and 0.3 m in wavelength or in energy about 10^{-6} aJ to 10^{-4} aJ.

Vibrational–rotational Transitions

What are usually called *vibrational transitions* involve transitions between the occupied rotational levels of the lowest vibrational state to rotational levels of the next highest vibrational state (transitions to higher ones are forbidden by selection rules and occur only very weakly). Thus the vibrational transitions are in fact a band of separate vibrational–rotational transitions. The change in vibrational level is the same for each one with different changes in rotational levels. Note that the scale of energy in Figure 7.1 is very much expanded for the rotational levels ($\sim 10^{-5}$ aJ) compared with that for the vibrational levels ($\sim$0.05 aJ). Sometimes the individual levels are resolved (molecules with few and light atoms), but generally the vibrational–rotational transitions occur as bands with little or no structure. The transitions occur in the region of about 200–4000 cm^{-1} or in energy units of about 0.004–0.08 aJ. The width of the bands is generally about 5 cm^{-1} or about 10^{-4} aJ.

Electronic Transitions

Electronic transitions also occur from the same rotational levels as in rotational and vibrational–rotational transitions and go to the rotational levels of many of the vibrational levels of a higher electronic state. Again, as for the vibrational–rotational, the rotational structure is often not resolved and sometimes not even the vibrational structure is resolved, so that electronic transitions occur as very broad (sometimes up to 7000 cm^{-1}) bands with varying amounts of structure. They occur in the range of about 10,000–50,000 cm^{-1} and beyond (0.2–1 aJ). A summary of the regions of the electromagnetic spectrum and the types and use of the spectroscopy involved is given in Table 7.1. We now discuss some of the details of the three different types of spectra associated with these three sets of energy-level changes.

Table 7.1 Regions of the electromagnetic spectrum and the spectroscopy involved

Region	Energy [aJ]	Wavelength [m]	Frequency [Hz]	Wavenumber [cm^{-1}]	Type of Spectroscopy	Source	Dispersion	Detector	Energy Levels Involved	Information Obtained
Radio Long-wave Short-wave TV	 2×10^{-13} 2×10^{-9} 2×10^{-7}	 10^{6} (=1 Mm) 10^{2} 1	 3×10^{2} 3×10^{6} 3×10^{7}		Nuclear magnetic resonance	Radio-frequency oscillator	Varying magnetic field	Radio-frequency coil (detect energy loss)	Nuclear spin levels (they have associated magnetic moments in different directions in a magnetic field)	For protons, no. and different kinds of protons in a molecule, i.e., qualitative analysis; similarly for other nuclei
Microwave	6×10^{-7} 2×10^{-5} 2×10^{-4}	0.3 0.01 0.001(=1 mm)	10^{9} 3×10^{10} 3×10^{11}	0.03 1 10	Rotational (microwave)	Klystron	Each sort of klystron can be varied over a small-frequency range	Crystal detector	Rotational—depend on masses and internuclear distances, and molecule must have a dipole moment	Geometries of gaseous molecules
Far infrared Infrared Near infrared	2×10^{-3} 0.02 0.2	10^{-4} 10^{-5} 10^{-6} (=1 μm)	3×10^{12} 3×10^{13} 3×10^{14}	100 1,000 10,000	Vibrational (infrared)	Glow from heated rare-earth oxides	Grating, prism (NaCl, KBr etc.)	Thermocouple	Vibrational–rotational bands	Some resolved lines can be analyzed to get geometries; others in particular locations serve to identify functional groups such as C=0, OH, etc.
Visible Ultraviolet	2.8 5 10	7×10^{-7} (=700 nm) 4×10^{-7} (=400 nm) 2×10^{-7} (=200 nm)	5×10^{14} 8×10^{14} 1.5×10^{15}	14,000 25,000 50,000	Electronic	Tungsten lamp H_2 or D_2 lamp	Prisms (glass and quartz), gratings	Photomultiplier, photographic plate	Electronic–vibrational–rotational bands often give rise to broad spectra	Some limited use for identifying functional groups can be used for quantitative analysis because absorption intensity can be accurately measured
Vacuum ultraviolet	20	10^{-7} (=100 nm)	3×10^{15}	100,000		Rare gas discharge				
X-Ray γ-Rays	2×10^{3} 2×10^{5}	10^{-9} 10^{-11}	3×10^{17} 3×10^{19}							

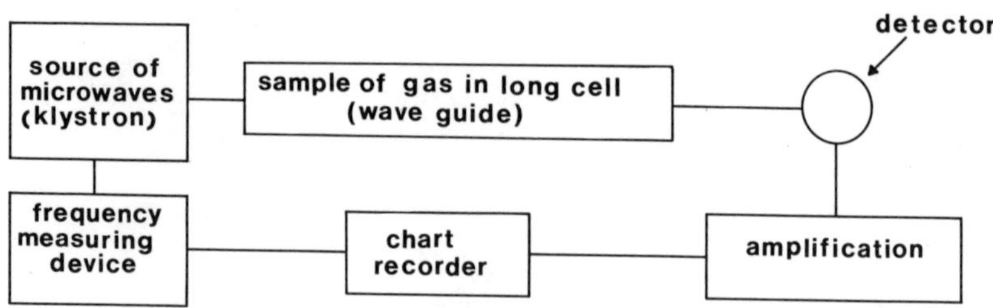

FIGURE 7.2. A schematic of a microwave spectrometer.

Microwave Spectroscopy

Spectroscopy involved in changes in rotational levels only is generally referred to as *microwave spectroscopy,* for the simple reason that this is the region of the electromagnetic spectrum in which the transitions occur. The units used to specify this radiation are given in Table 7.1.

We measure the microwave spectrum of a molecule by passing microwave radiation through a gaseous sample (the only phase in which molecules rotate completely freely), steadily varying the frequency of the radiation and seeing which frequencies are absorbed by the gas. The general layout of the equipment is shown in Figure 7.2.

The absorption spectrum of the sample is obtained on the chart recorder when the variation in signal intensity is plotted as a function of frequency. Typically the spectrum consists of a number of extremely sharp lines well separated from one another. An example is given in Figure 7.3. Carbonyl sulfide and other linear molecules give rather simple rotational spectra. Nonlinear molecules often give very complicated spectra. However, it is possible from a detailed study of the absorbed frequencies to work back and find the precise details of the rotational energy levels of a molecule.

FIGURE 7.3. Microwave spectrum of carbonyl sulfide.

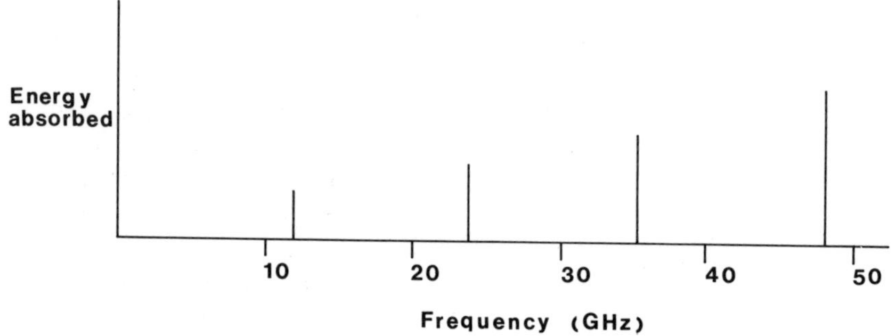

We now want to see what details of molecular structure can be obtained from a precise knowledge of the molecular rotational energy levels. Classically, the kinetic energy of rotation of a body about some axis is

$$T = \frac{P^2}{2I} \tag{7.3}$$

where P is the angular momentum about the axis and I is the moment of inertia about the same axis. The rotational motions of two molecules, OCS and H_2O, about different axes are illustrated in Figure 7.4. Each atom in the molecule makes a contribution to the moment of inertia of $m_i r_i^2$, where m_i is the mass of the atom and r_i is its distance from the axis. The angular momentum is quantized and hence the energy levels (described by a quantum number $J = 0, 1, 2, 3, 4, \ldots$) are discrete. The experimental location of these levels leads to values of I about different axes in a molecule (about only one axis in a diatomic molecule) and thus to the r_i values. As a result, precise values for the molecular geometry are obtained. At present microwave spectroscopy is the most precise method available for the determination of geometries. However, it must be realized that this method is restricted to compounds that have an appreciable vapor pressure. Furthermore, selection rules that specify conditions to be met when radiation is absorbed restrict its applicability to molecules that have a permanent dipole moment. There is no microwave absorption for molecules such as H_2, Cl_2, CO_2, or benzene. The transitions between their rotational levels are forbidden by the dipole-moment selection rule.

Infrared Spectroscopy

When absorption or emission of radiation occurs due primarily to changes in vibrational energy levels of molecules, the relevant region of the electromagnetic spectrum is the infrared, typically about 200–4000 cm^{-1}. The other equivalent units are given in Table 7.1.

Again quantum mechanics predicts that vibrational energy levels in molecules are quantized and hence discrete. In this case the difference between energy levels for allowed transitions corresponds to the frequency of the classical vibrational motion of the molecule where each atom oscillates about its equilibrium position with a simple harmonic motion. Figure 7.5 shows the vibrational motions of a few simple molecules. In applying infrared spectroscopy to qualitative analysis, the jargon used is illustrated in the simplest cases in Figure 7.5. Notice that

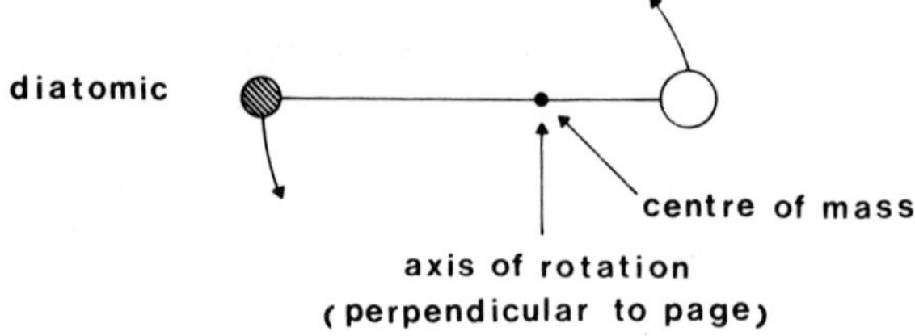

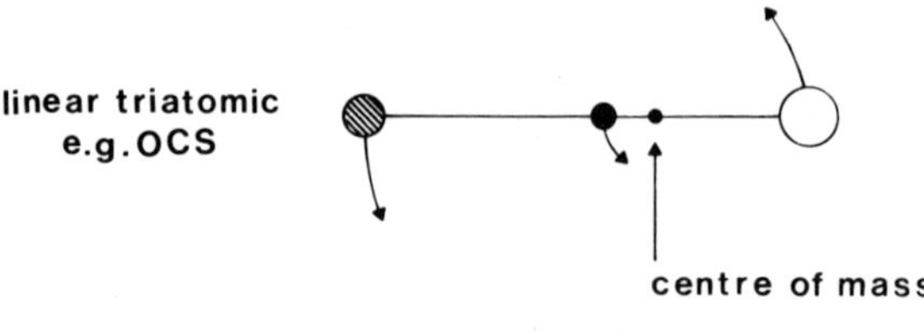

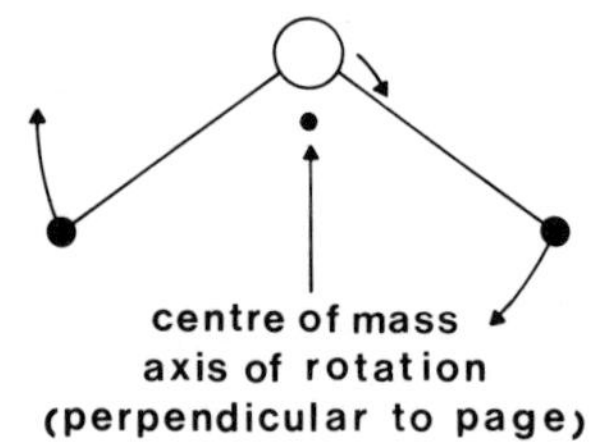

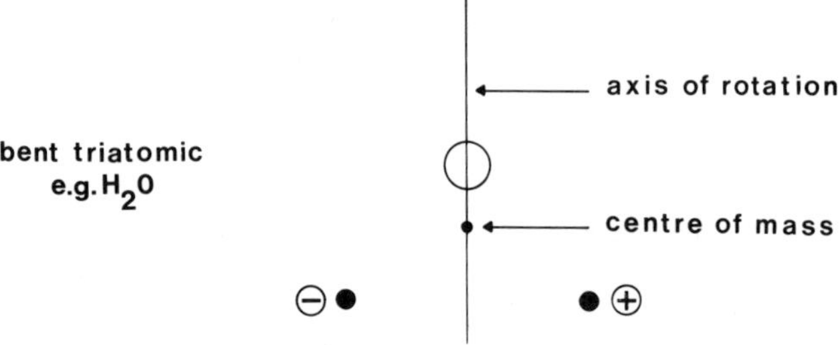

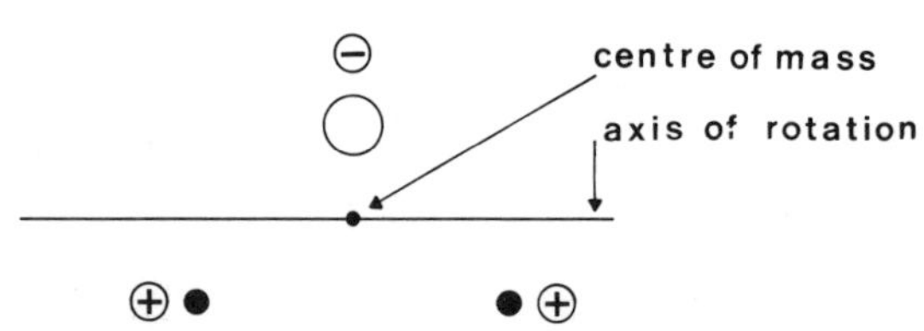

FIGURE 7.4. Rotational motions of molecules, for which the energies are quantized. ⊕ and ⊖ mean motion out from or into the page.

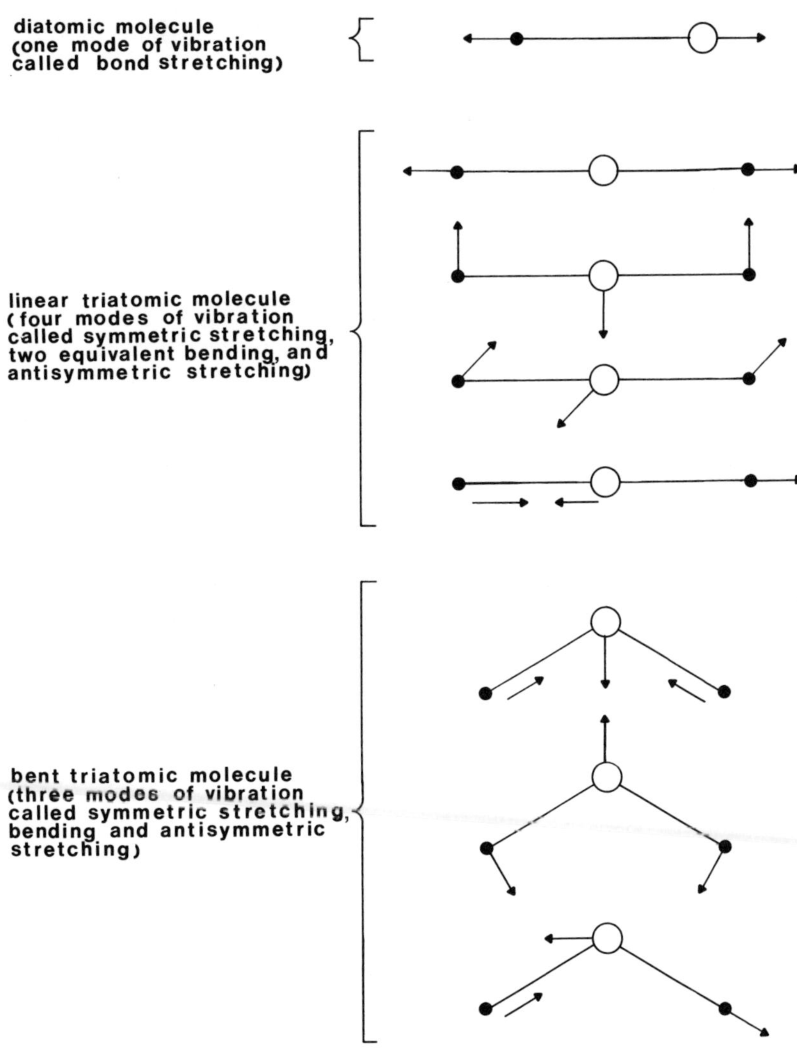

FIGURE 7.5. Vibrational motions of simple molecules. The circles represent the nuclei of the molecule.

the terms for each type of vibration are related to the sort of motion performed by the nuclei in the molecule. They are *symmetric bond stretching, antisymmetric bond stretching* (this will only occur when there are two equivalent bonds), and *angle bending*. In molecules with many more atoms than three, although all of the atoms in the molecule are strictly involved in the vibrations, it turns out that to a good approximation many of the vibrations are located in one (or a few) particular chemical bond(s). For example, in organic aldehydes and ketones there is always a vibration called C=O stretching, in which the main motion in the molecule is extending and compressing the C=O bond. Because of this and because the vibration frequency is dependent on the strength of the bond, this vibration and hence its absorption in an infrared spectrum occurs at about the same frequency regardless of the other parts of the molecule. The small variations that occur can then be useful as a guide to deciding the other groups in the molecule. For example, this C=O stretching vibration always occurs at 1710–1720 cm^{-1} for aliphatic ketones, but if the C=O group is conjugated in some way such as in aromatic ketones the frequency is lowered by 20–60 cm^{-1}, depending on the exact nature of the conjugation.

Other vibrations in large molecules correspond to stretching and bending of the various groups in the molecule. There are some vibrations that correspond to motion in which all the atoms in the molecule move to some extent. These so-called skeletal vibrations are not as useful for identification purposes.

The great advantage of infrared spectroscopy is that each different kind of bond has a different *characteristic group frequency* (a designation that could have been used for the C=O stretching vibration). These can often be recognized in the infrared spectrum of the compound and so provide some information on the structure of the molecule.

The infrared spectra of SO_2 and acetaldehyde are shown in Figure 7.6. For SO_2 the three regions of absorption correspond to the three vibrations shown for the bent molecule in Figure 7.5. The first one has been assigned to the band at 1152 cm^{-1}, the second to that at 517 cm^{-1}, and the third to that at 1362 cm^{-1}. The numbers for these vibrations illustrate two general trends for vibration frequencies of groups that help in locating them. The antisymmetric stretching frequency is generally higher than the symmetric stretching frequency (naturally for the same pair of bonds), and the bending frequency (again for the same system) is about half the stretching frequency.

For acetaldehyde a number of the bands in the spectrum can be assigned to group frequencies. These assignments are often made by a careful comparison of the spectra of a series of closely related molecules. The assignments are:

2960 cm^{-1}—antisymmetric C—H stretch (CH_3 group)

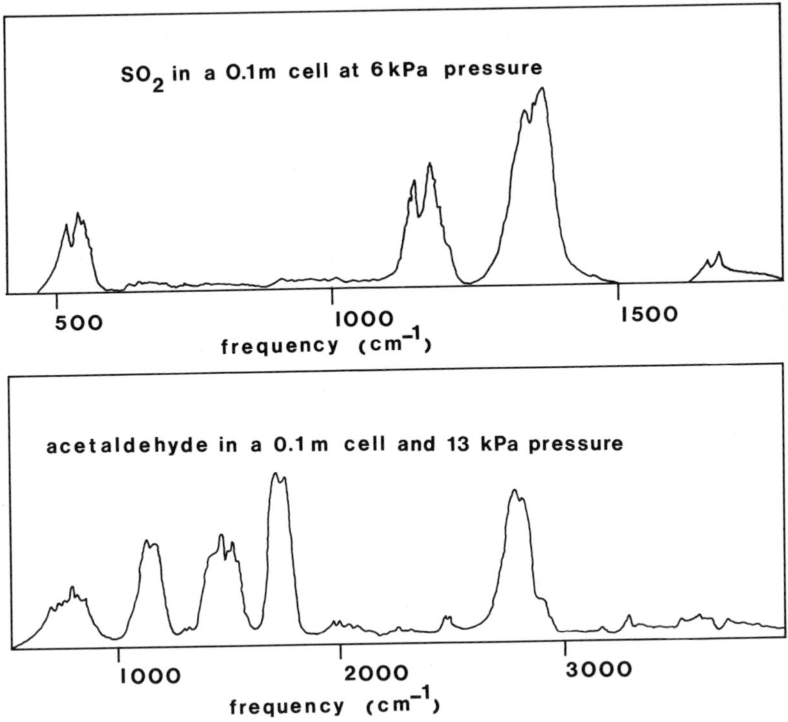

FIGURE 7.6. Infrared spectra of SO_2 (gas) and acetaldehyde (gas).

2870 cm^{-1}—symmetric C—H stretch (CH_3 group)
2820 cm^{-1}—C—H stretch (aldehyde C—H)
1725 cm^{-1}—C═O stretch
1380–1460 cm^{-1}—a number of C—H bending vibrations
1100 cm^{-1}—C—C stretch.

From studies of the infrared spectra of many compounds, tables of characteristic group frequencies have been prepared. A small selection of frequencies for the more common organic compounds (containing C, H, O, N, and the halogens) is given in Figure 7.7. Notice that for some vibrations quite wide frequency ranges are indicated. This means, as mentioned previously, that variations in the exact frequency can often be correlated with the neighboring substituents. It should also be remembered that the identification of functional groups by infrared spectroscopy is not limited to simple organic compounds. The technique is useful for all classes of compounds.

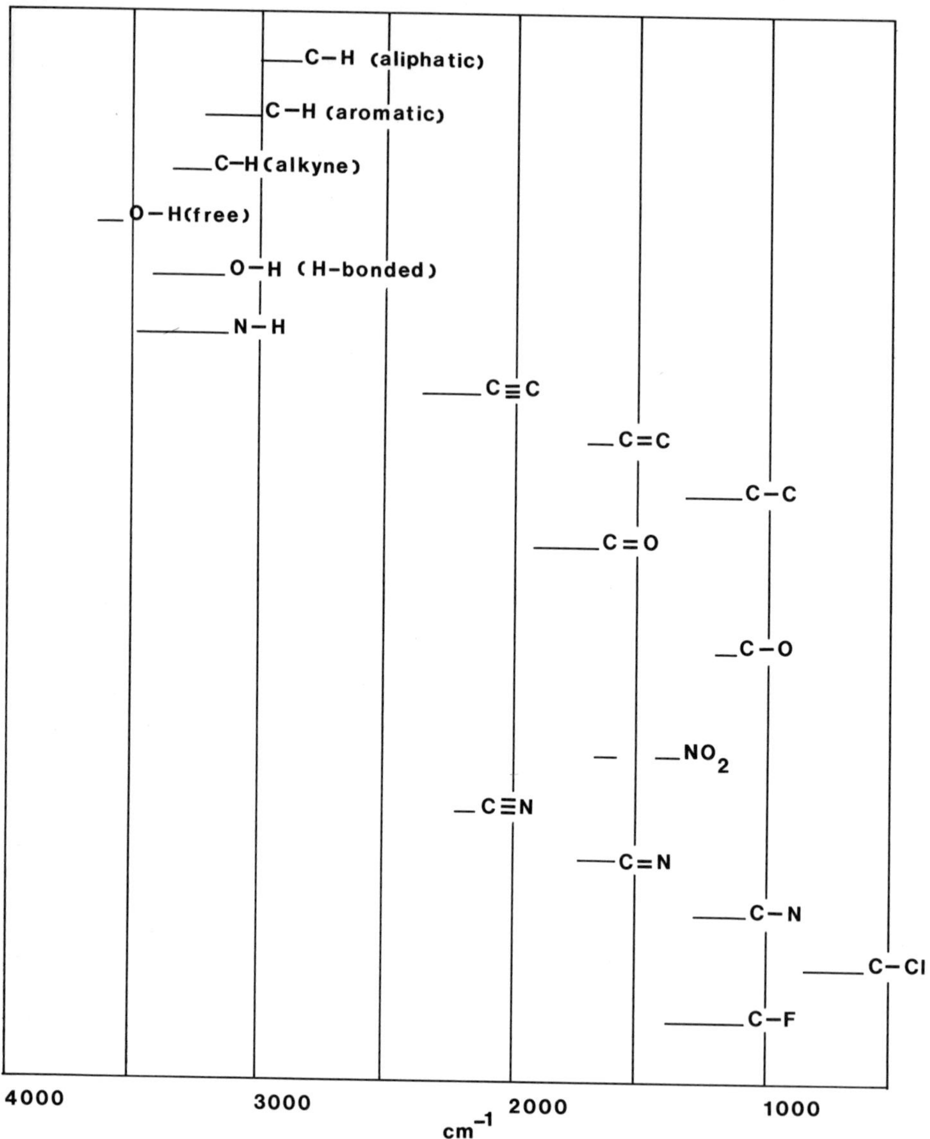

FIGURE 7.7. Characteristic group frequencies for stretching vibrations in common organic molecules.

The basic features of an infrared spectrometer are shown in Figure 7.8. This is a double-beam instrument, the essence of which is to subtract out absorption due to solvent and cell windows and normalize the percentage absorption at each wavelength. The light from S is split into two equal beams by M1 and focused by SM1 and SM2 to pass through the sample and reference cells SC and RC.

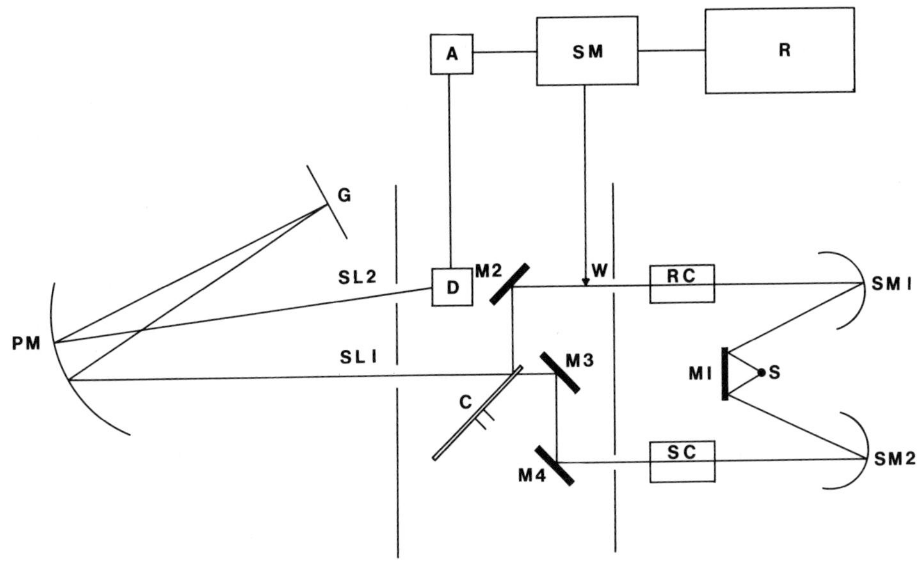

FIGURE 7.8. Basic features of a double-beam infrared spectrometer with: S—infrared light source (Nerst glower); M1, M2, M3, M4—plane mirrors; SM1, SM2—spherical mirrors; SC—sample cell; RC—reference cell; W—optical wedge; C—semicircular chopper; PM—paraboloidal mirror; G—diffraction grating; D—detector; A—amplifier; SM—servomotor; R—chart recorder; SL1, SL2—entrance and exit slits.

The light beams are combined by use of the mirrors M2, M3, and M4 and the semicircular mirror on the chopper, that alternatively passes through light from the sample side and the reference side. This beam then goes through the entrance slit SL1 and via the paraboloidal mirror PM to the grating, which is driven to select the different wavelengths going through the exit slit SL2 to the detector. The amplified signal drives a servomotor, SM, which moves to balance the beams by the wedge W. This movement of the wedge is proportional to the amount of light absorbed in the sample cell relative to the reference cell. Hence we obtain a graph on the recorder of absorption versus wavelength.

Most infrared studies for qualitative analysis are made on liquids and solids. For liquids a thin film (say, 0.01 mm thick) between two plates of solid sodium chloride (which is transparent to infrared radiation down to about 400 cm^{-1}, in contrast to glass or silica, which have strong absorption in the infrared) is usually employed. The other common method is to use solutions generally with cells of somewhat longer pathlength. Care must be exercised in choosing the solvent because most solvents absorb in some region of the infrared. Two common solvents with relatively small regions of absorption are CS_2 and CCl_4. CS_2 is unsuitable at 1400–1600 cm^{-1} and 2000–2250 cm^{-1}, while CCl_4 is unsuitable at 700–850 cm^{-1}. For those solids for which it is not feasible to make up a solution since they are very sparingly soluble, a fine suspension in paraffin oil (Nujol), often referred to as a *Nujol mull,* is frequently used. Gases can also be studied, with the use of cells having a pathlength of 0.1 m and sometimes longer.

Small molecules studied at low gas pressures often show rotational fine structure. When this is so, it is often possible to analyze the fine structure and derive moments of inertia and hence geometries, the computations being similar to those in microwave spectroscopy. For example, the geometry of CO_2 has been accurately determined in this way.

Visible-ultraviolet Spectroscopy

As with microwave and infrared spectroscopy, visible-ultraviolet or electronic spectroscopy is mostly studied by adsorption methods, although here the emission experiments are somewhat easier to perform than in the infrared and microwave regions. The general layout of the instrumentation is similar to that of infrared spectrometers. Of course the light sources will be different, generally with a tungsten lamp for the visible and a hydrogen or deuterium discharge lamp for the ultraviolet. The optical material for the cells and the means of dispersion, if a prism is used, will be quartz (ultraviolet) or glass (visible), rather than sodium chloride. Quite often gratings are used as in infrared spectroscopy.

Electronic absorption spectra are usually studied on dilute solutions (often 10^{-3} to 10^{-5} mol dm^{-3}) in water or other suitable transparent solvents in cells that vary from 0.1 mm to about 0.1 m thick. When vapors are studied the cells may be of greater length if the vapor pressure is kept low enough.

As mentioned previously, the transitions involved in electronic spectroscopy involve changes in rotational and vibrational levels as well as a change in electronic levels, but often the fine structure is washed out. When fine structure is present it is possible to analyze the spectra and obtain some geometric structural information, In general this happens only for small or highly symmetrical molecules. For most molecules the electronic absorption spectra are broad and fairly structureless. However, they are still useful for qualitative analysis and for quantitative analysis of a given known compound.

Electronic transitions associated with absorption of organic molecules in the visible or ultraviolet are usually associated with the presence of one or more double or triple bonds. Such structural features that lead to absorption are collectively known as *chromophores*. The change in electronic state associated with the chromophore is generally one of two types. In the first type an electron is promoted from a bonding π-orbital to an antibonding π^*-orbital. Here the associated transition is called a "$\pi^* \leftarrow \pi$" transition and usually has a high intensity (high ϵ value; see later). The second occurs when there is an atom in a double-bonded system with a lone pair, such as—C═O. Then besides having a $\pi^* \leftarrow \pi$ transition, there is another transition in which an electron is promoted from one of the lone-pair (nonbonding) orbitals to the π^*-orbital. This is called an "$\pi^* \leftarrow n$" transition and differs from the $\pi^* \leftarrow \pi$ transition in that it is usually of longer wavelength (lower energy) and lower intensity (low ϵ value). The transition is formally forbidden by the selection rules and gains its intensity from small perturbing effects.

Figure 7.9 shows the absorption spectrum of acetaldehyde, which shows the $\pi^* \leftarrow n$ transition with maximum at about 35,000 cm^{-1}

FIGURE 7.9. Ultraviolet absorption spectrum of acetaldehyde (gas in 0.1-m cell at a pressure of 13-kPa).

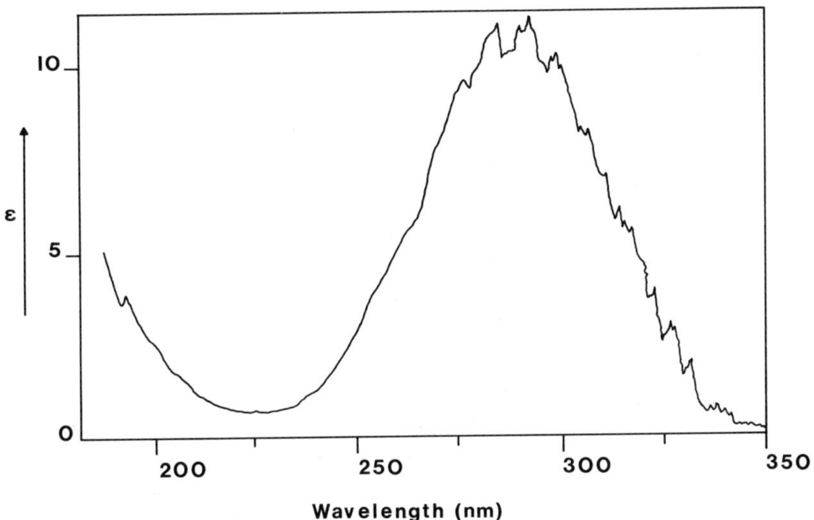

FIGURE 7.10. Beer's law.

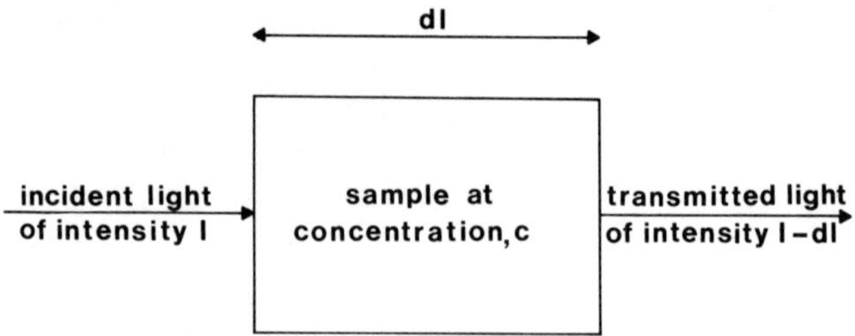

Visible and ultraviolet spectra are not as useful as structural probes as are other spectroscopic techniques but can be used extensively for quantitative measurements of concentrations. As indicated previously, this is measured by a quantity ϵ, called the *molar extinction coefficient,* which is the proportionality factor in Beer's law for a particular transition. Beer's law states that the decrease in intensity of a light beam as it passes through an absorbing medium is proportional to the original intensity, the concentration of the solution, and the distance through which it passed. If we consider the system illustrated in Figure 7.10, the loss of intensity ($-dI$) will be proportional to c (the concentration in mol dm^{-3}), dl (the distance in cm through which the light passes),[1] and the intensity of the beam, I:

$$-dI = kIc\,dl. \tag{7.4}$$

On integrating we obtain

$$-\ln I = kcl + \text{const.} \tag{7.5}$$

Then if the initial light intensity is I_0 we write

$$\ln \frac{I_0}{I} = kcl \tag{7.6}$$

or

$$\log_{10} \frac{I_0}{I} = \epsilon cl \tag{7.7}$$

where

$$\epsilon = \frac{k}{2.303}$$

and is the quantity called the *molar extinction coefficient;* $\log_{10}(I_0/I)$ is often called the *absorbance* or *optical density* and given the symbol A:

$$A = \log_{10} \frac{I_0}{I} = \epsilon cl. \tag{7.8}$$

[1]Notice that ϵ is not in conventional SI units; the units used are equal to 10^{-1} mol^{-1} m^2.

Thus if we use solutions of known concentration, we can initially determine ϵ at a chosen wavelength, and then determine the concentrations of other solutions by measuring their absorbances and using Equation 7.8. The spectrum of acetaldehyde in Figure 7.9 is shown as a plot of ϵ versus wavelength. This is the commonest way of presenting visible and ultraviolet spectra. The wavelengths are now quoted in nm but in the earlier literature non-SI units such as Å or mμ are found (1 nm = 1 mμ = 10 Å).

Nuclear Magnetic Resonance Spectroscopy

Besides observing transitions between the energy levels of molecules illustrated in Figure 7.1, there are various methods based on separating degenerate energy levels (different states with the same energy) by applying a magnetic field. One of these has been particularly useful in chemistry and is known as *nuclear magnetic resonance spectroscopy,* or simply NMR. It depends on the fact that some atomic nuclei possess a nuclear spin and hence have a magnetic moment, such that the nucleus behaves like a tiny bar magnet. When a magnet is in a magnetic field its energy depends on the direction in which it points relative to the direction of the magnetic field. Thus a compass needle settles down pointing north, the direction of lowest energy. In any other direction the energy is higher. Pointing in the opposite direction is the higher energy configuation and how much higher it is depends directly on the strength of the magnetic field.

In the case of an atomic nucleus that possesses nuclear spin, the energy of orientation in a magnetic field is quantized in a manner similar to the quantization of electron spin. For example, the proton, which has a nuclear spin (described by a quantum number I) of ½ (cf. $s = ½$ for an electron), has two directions in which the energy is quantized in a magnetic field [described by the quantum number $M_I = ½$ or $-½$ (cf. $m_s = ½$ or $-½$ for an electron)]. These two orientations correspond to a tiny energy difference when the proton is in a very large magnetic field and, as for a magnet, this difference varies with the strength of the field. For example, when the field is about 2.35 T the energy difference corresponds to around 100 MHz, and when it is 1.4 T the difference is around 60 MHz. If radiation of the appropriate frequency is supplied, it can cause transitions between the two levels and energy is absorbed—this is NMR absorption.

It is possible to determine an NMR absorption spectrum by applying a steady magnetic field and irradiating the sample with radiation while slowly varying its frequency. However, an alternative that happens to be more convenient experimentally is to slowly vary the magnetic field strength while using radiation of an accurately fixed frequency. Modern NMR spectrometers operate in this way. Typically a sample of the compound in a small glass tube is spinning within a coil connected to a crystal-stabilized 100-MHz radiofrequency oscillator, the coil being located between the poles of a large electromagnet. Sweep coils are wound around the poles to enable the magnetic field to be varied slowly. When the field is about 2.35 T the detector (see Figure 7.11) that is linked to the radiofrequency oscillator detects absorption of power by the sample, corresponding to the NMR absorption by protons. The magnetic field can be varied by a small amount, but sufficient to cover the region where different sorts of protons absorb (see Figure 7.11).

The importance of NMR spectroscopy for structural studies stems from the fact that the valence electrons in a molecule containing protons produce tiny magnetic fields because of the electron circulation in the molecule. The electrons thus provide a small degree of shielding or deshielding of the protons from the applied magnetic field; that is, they experience a field slightly less or greater than

FIGURE 7.11. Basic features of an NMR spectrometer.

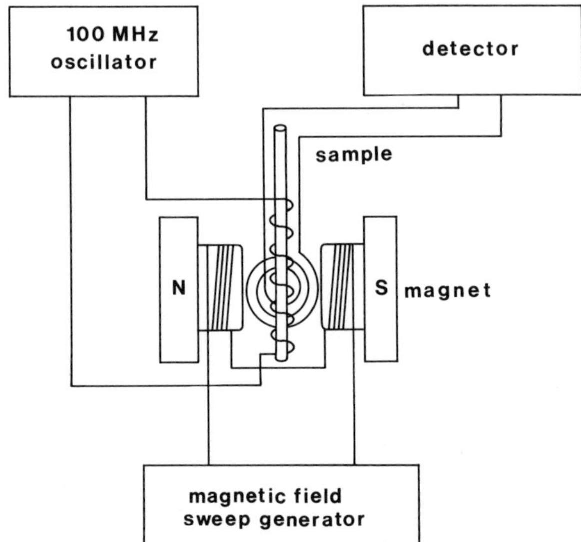

the applied field. How great this is depends on the electronic environment of the individual protons. The electronic environment varies with each type of proton in a molecule, so that the absorption occurs at slightly different places in the spectrum. For example, in very pure ethanol there are three regions of absorption. As can be seen from the NMR spectrum of ethanol in Figure 7.12, the amount of absorption (area under peaks) for the three different kinds of proton is in the ratio of 1:2:3, corresponding to the ratio of the number of different kinds of protons. Nuclear magnetic resonance spectra are run with an added standard, tetramethyl silane (TMS), and all absorptions are measured relative to this. On the abscissa is the chemical-shift parameter δ; "chemical shift" means how far the absorption is shifted relative to TMS and is given (to remove the magnetic-field dependence) by

$$\delta = \frac{B_{\text{sample}} - B_{\text{TMS}}}{B_{\text{TMS}}} \times 10^6 \tag{7.9}$$

where B means the magnetic field strength at which the absorption occurs. The unit, parts per million (ppm), comes from multiplying the ratio by 10^6. It should be noted that in a molecule such as ethanol free rotation about single bonds is sufficiently rapid that all the methyl protons are in identical chemical environments and hence have the

FIGURE 7.12. Nuclear magnetic resonance absorption of ethanol.

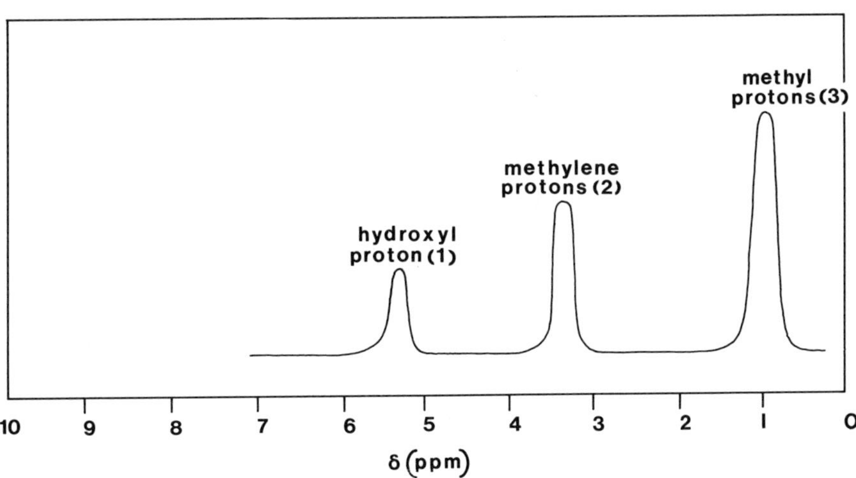

same chemical shift. The same applies to the methylene protons.

As another example the NMR spectrum of acetaldehyde is shown in Figure 7.13. Notice in Figure 7.13 that the methyl protons absorb at a value of δ different from that where the methyl protons of ethanol absorb. This is because in the first case there is $HOCH_2$—next to the methyl group and in the second,—CHO. This gives the two methyl groups slightly different electronic environments. Hence we see that NMR spectroscopy is a very sensitive probe of this and is of great use in qualitative analysis.

The nuclei ^{12}C and ^{16}O, the most abundant isotopes of C and O, have zero nuclear spin and hence show no NMR absorptions. ^{13}C, however, has a nuclear spin of ½ and shows NMR absorptions, but at quite a different magnetic field strength from that appropriate for protons. However, the isotopic abundance is very low ($\approx$1%), and very sensitive techniques have to be used to detect ^{13}C NMR. This technique has developed rapidly in the last few years and promises to be a very powerful method of structural analysis. Of course there are many other nuclei with nuclear spins that are not zero. Two of the commonest besides ^{1}H and ^{13}C, that have been used in NMR are ^{19}F and ^{31}P. Again, they occur at very different field strengths from those of other nuclei.

Just as for infrared spectra, tables of chemical shifts can be useful for structural analysis. Figure 7.14 is an abbreviated table for proton chemical shifts of compounds containing C, H, and O.

Figure 7.13. Nuclear magnetic resonance spectrum of acetaldehyde.

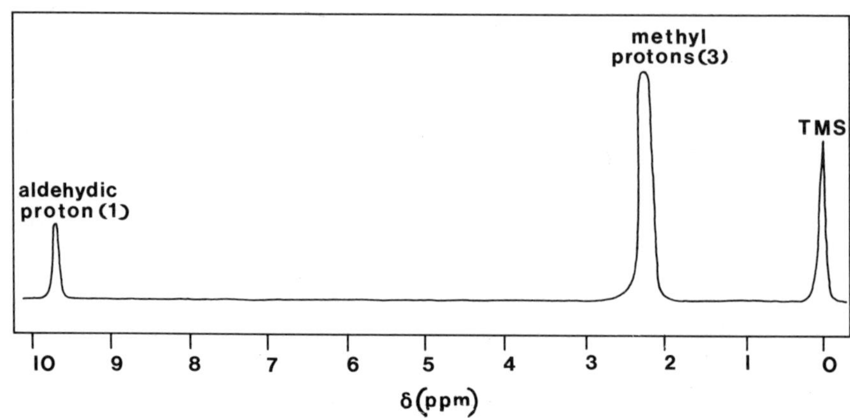

Additional information can be obtained from NMR spectra run at high resolution. Absorption bands due to different sets of chemically equivalent nuclei are often further split into a number of closely spaced lines, due to very small interactions with neighboring nuclei. This is referred to as *spin–spin splitting* and can often be used to determine the arrangement of different groups in a molecule.

Mass Spectroscopy

Mass spectroscopy is so named because we detect a spectrum of mass-to-charge quotients for the various ions formed when a molecule is ionized by electron bombardment and the molecular ion also breaks up into smaller charged fragments. From the spectrum we obtain two useful pieces of information: (a) the accurate molecular weight of the molecule and (b) some knowledge of the molecular structure by examining which particular fragment ions are formed.

FIGURE 7.14. Proton chemical shifts (δ) for a few common organic groups.

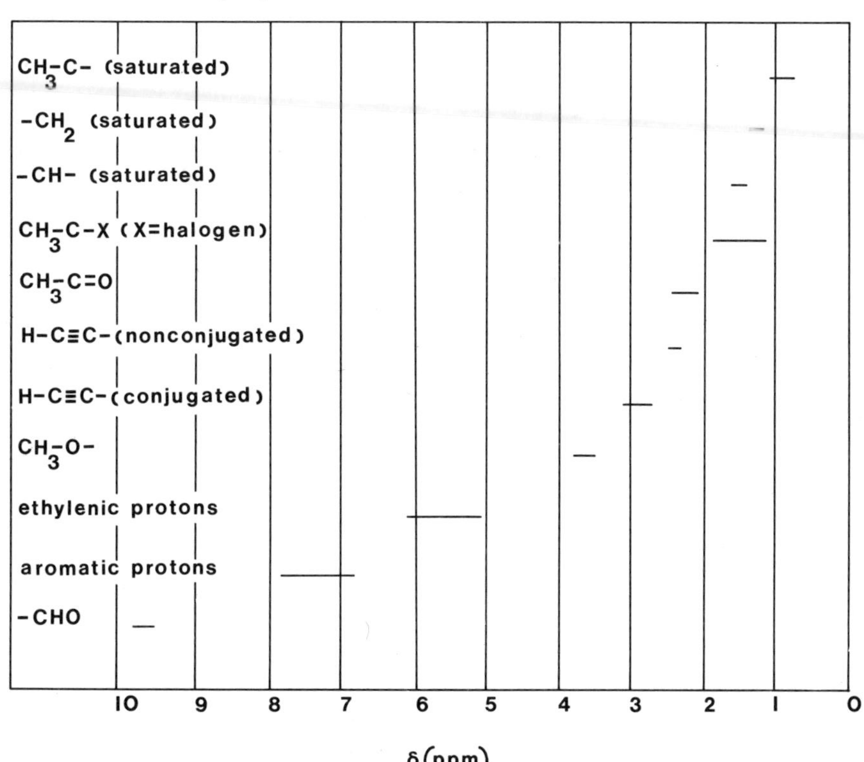

The mass spectrometer consists of five main components, as illustrated in Figure 7.15. A description of these is as follows:

1. Some means of volatilizing the sample[2] and getting it into a very high vacuum chamber ($\sim 10^{-4}$Nm^{-2}). Only about 50 μg of sample is sufficient to run a spectrum.
2. A chamber where the sample is bombarded with high-energy electrons ($\sim$70 eV is typical). This process forms the molecular ion, M^+. Since a typical energy necessary for this is of the order of 10 eV, M^+ has enough excess energy to break bonds and form a variety of fragment ions.
3. An accelerating potential region where the various positive ions are accelerated and travel with high velocity into the analyzing portion of the spectrometer.
4. The analyzing part of the instrument involves separation of the various ions of differing mass-to-charge quotient[3] (m/e) in a strong magnetic field. For high resolution, the ion beam is first made monoenergetic by an electrostatic analyzer (not shown in Figure 7.15) before it is passed through the magnetic field. The

[2]The very extensive use of mass spectroscopy in the last decade or so is partly a result of the development of techniques to volatilize even very highly involatile substances in the amounts necessary to run a mass spectrum.

[3]Strictly speaking, we should write m/ne, n being the number of electrons lost by the neutral molecule or fragment. Sometimes n is greater than unity.

FIGURE 7.15. Schematic of a mass spectrometer.

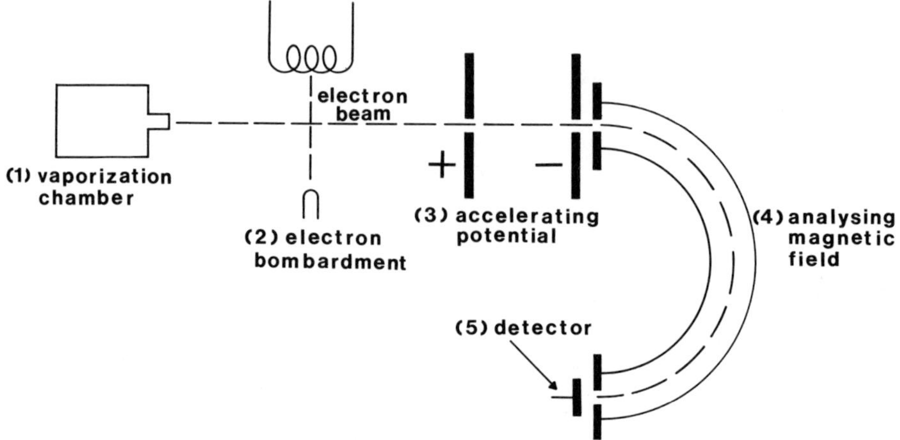

magnetic field deflects ions of large m/e less than those of smaller m/e. The radius of the circular path that a given ion travels is found by equating the centripetal force to the force exerted on the charged ion by the magnetic field:

$$\frac{mv^2}{r} = Bnev$$

and since
$$\tfrac{1}{2}\, mv^2 = neV \qquad (7.10)$$
$$\frac{m}{ne} = \frac{r^2B^2}{2V}$$

where B is the magnetic field strength and V is the accelerating potential.

5. The detector, which has some means of detecting the number of ions of a given m/e that hit it. From Equation 7.10 it can be seen the spectrum can be scanned by varying either B or V. The end result is a mass spectrum that shows the relative abundance of ions of varying m/e. As an example, the mass spectrum of acetaldehyde is shown in Figure 7.16.

Notice that the most intense peak has $m/e = 29$, probably due to the fragment ion CHO^+. The molecular ion is at m/e 44, $C_2H_4O^+$ at m/e 43, and the peak at m/e 15 corresponds to CH_3^+. Notice that a peak occurs at m/e 18 due to water impurity. For more complicated molecules one needs some familiarity with the factors controlling general fragmentation modes to deduce structural information. This often gives many clues to structures, but of course they cannot always be solved completely.

FIGURE 7.16. Mass spectrum of acetaldehyde.

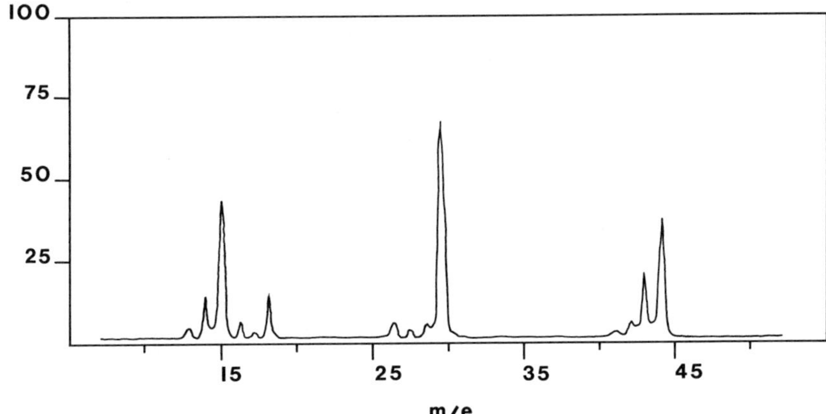

Many peaks have ambiguous assignments unless the spectra are recorded at very high resolution. For example, CO^+, H_2CN^+, N_2^+, and $C_2H_4^+$ show molecular ions at *m/e* 28 under low resolution. Under high resolution conditions these occur at 27.9949, 28.0187, 28.0313, and 28.0061 and thus can be distinguished.

Diffraction Techniques

Diffraction techniques involve careful measurements of the way that a beam of light or some other wavelike phenomenon is affected by passing through, or being reflected from, matter. We can gain some understanding of the basic principle of diffraction techniques by considering what happens when a beam of monochromatic light passes through an opaque plate with a number of very narrow, parallel slits cut in it (a diffraction grating). In certain directions on the far side of the grating the diffracted light will be more intense than in others (Figure 7.17).

If we consider a point P (Figure 7.17) at which two waves arrive in phase and reinforce each other, we have:

$$S_1P - S_2P = (5\lambda - 4\lambda) = n\lambda, \qquad (n = 1). \qquad (7.11)$$

For similar points along the second, third, fourth order lines n will be 2, 3, and 4, respectively. If the point P is chosen sufficiently far away from the grating, the lines S_1P and S_2P become nearly parallel (Figure 7.17). The path difference AS_1 is a function of d and the angle θ. For parallel paths the angle S_1AS_2 is 90°, and hence

$$AS_1 = S_1P - S_2P = d \sin \theta. \qquad (7.12)$$

Thus from Equation 7.12 we get

$$n\lambda = d \sin \theta. \qquad (7.13)$$

For monochromatic light, that is, with wavelength (λ) fixed at a constant value, there will be a series of beams of maximum intensity at different values of θ corresponding to $n = 0, 1, 2, 3, \ldots$. Thus if we know λ and measure the first few values of θ (so that we are sure of the correct values to assign to n), we can use Equation 7.12 to calculate d, that is the spacing of the slits. We might notice that to obtain readily measurable values of the diffraction angle θ corresponding to small values of n it is necessary for the wavelength of light to be similar to the spacing of the slits, d.

In a crystalline solid we have atoms packed in a regular manner and able to diffract light in a way similar to the slits described above. The spacings are so tiny (typically 100–

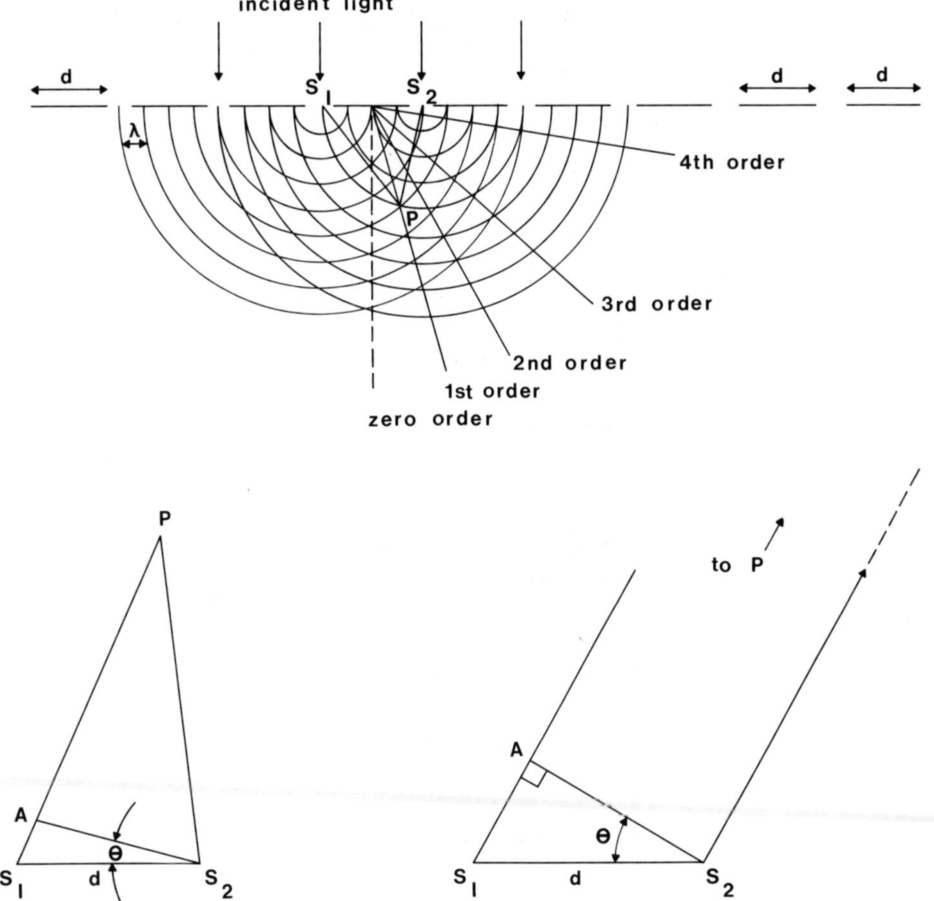

FIGURE 7.17. Illustrating the law of diffraction $n\lambda = d \sin \theta$.

200 pm) that light of very short wavelengths, such as X-rays, have to be used to ensure that λ and d are of similar magnitude (e.g., copper $K\alpha$ radiation, $\lambda = 154$ pm). Thus a study of *X-ray diffraction* by crystals makes it possible to determine the spacings of atoms in the crystal. Of course the situation is more complicated than the case of the simple diffraction grating described above. There will usually be several different spacings within the crystal and several types of atoms, each with a different capacity for diffracting X-rays. This renders the diffraction pattern very complicated, and the process of calculating all of the relevant spacings in the crystal is usually a formidable task. However, the principle is still essentially the same as for the simple grating.

By means of the X-ray-diffraction technique the structures of many thousands of crystalline materials have been elucidated so that we now know the precise relative positions of all of the atoms within the crystal.

A small, single crystal of the substance is normally used in X-ray-diffraction studies. If the atoms are not all regularly spaced, that is if the crystal is not perfect, the diffraction pattern is not so sharply defined. Thus the degree of crystallinity of a substance can be determined from the appearance of its diffraction pattern. Many so-called amorphous solids—glasses, resins, and so on—have been studied by this technique. Sometimes a finely powdered crystalline sample is deliberately used to obtain limited information about the sample. The diffraction pattern becomes a series of concentric rings around the X-ray-beam direction in this case.

One of the disadvantages of X-ray diffraction is that the power of an atom to diffract X-rays is proportional to its atomic number, and so the diffraction effect of light atoms is often swamped by the effects coming from heavier atoms. This is particularly troublesome for hydrogen, so much so that for many compounds it has not proved possible to determine the positions of hydrogen atoms within the crystal with acceptable accuracy.

This has led to the development of *neutron diffraction*. A beam of neutrons emerging from an atomic reactor at thermal energies has a wavelength, according to the de Broglie formula ($\lambda = h/p$), of around 150 pm. Hence this method is entirely suitable for studying crystal structures. Moreover, the diffracting effects of different atoms are all much the same (the overall variation among atoms is no more than a factor of 2 or 3) and so it is valuable for locating the positions of the light atoms and especially hydrogen.

The wave nature of *electrons* is also useful for diffraction studies; for example, the wavelength of 40-kV electrons is about 6 pm. However, electrons can penetrate only a very thin layer of solid, and so electron diffraction is more suitable for the study of surface structures. It is also suitable for studying diffraction by molecules in the gas phase. Indeed, much of our information on the geometries of small molecules has been obtained by electron-diffraction experiments. The focused electron beam in Figure 7.18 is passed through a thin stream of the sample gas, which is "squirted" across the electron beam from a nozzle and condensed on the other side. (If the gas were to diffuse through the whole apparatus, it would effectively

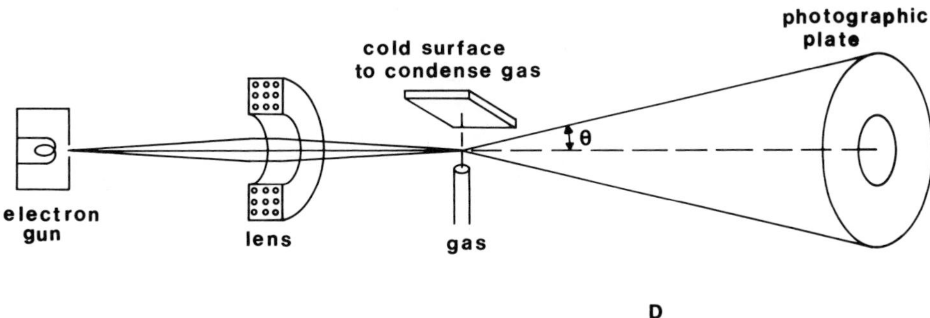

FIGURE 7.18. Electron diffraction apparatus.

absorb the electron beam and prevent the diffracted beam from reaching the photographic plate). The effect is rather like that of using a powdered sample in X-ray diffraction—the diffraction pattern is a series of concentric rings. By measuring the diameters of these and knowing the geometry of the apparatus and the wavelength it is possible to evaluate the distances between various pairs of nuclei in the molecule. For molecules that are not too large, that is, containing a relatively small number of nuclei, we finally derive the molecular structure from all of the internuclear distances. Again the diffracting power of the atom increases with its size so that when heavier atoms are present with hydrogen atoms, the latter may not be very accurately located. The electron-diffraction method has provided accurate data on molecular geometries for many small molecules and for some, such as SF_6 or CF_4, it is at present the only practicable technique for structure determination.

Table 7.2 summarizes the use and limitations of the diffraction techniques.

Table 7.2 Diffraction techniques, uses, and limitations

Diffraction technique	Phase of sample	Limitations
X-ray	Solid	Difficult to locate light atoms — X-rays scattered by electrons
Neutron	Solid	Detects light atoms because neutrons scattered by nuclei
Electron	Gas	Scattering weak—difficult to get very accurate geometries—useful for nonpolar molecules

Problems

7.1 Name the experimental technique most suitable for determining the following:

a. the C=C bond length in gaseous ethylene
b. the C—C bond length in diamond
c. the H—Cl bond length in gaseous HCl
d. whether a compound of empirical formula C_5H_{12} is

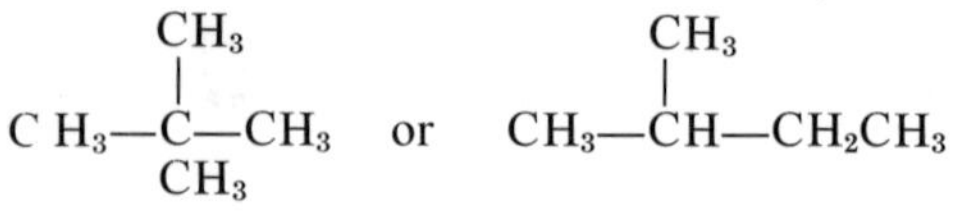

$$\begin{array}{c} CH_3 \\ | \\ CH_3\text{—}C\text{—}CH_3 \\ CH_3 \end{array} \quad \text{or} \quad \begin{array}{c} CH_3 \\ | \\ CH_3\text{—}CH\text{—}CH_2CH_3 \end{array}$$

e. Whether the CO bond in a molecule is a single or double bond.
f. The H—F distance in crystalline $K^+ (HF_2)^-$.
g. In a compound C_nH_{2n+2} whether $n = 3$ or 4.

Appendix I
Angular Momentum and the Bohr Atom

Angular momentum is defined classically in terms of the product of the *moment of inertia* I and the *angular velocity* ω.

$$L = I\omega \tag{A.1.1}$$

where these quantities are defined as

$$I = \sum_{i}^{n} m_i r_i^2 \tag{A.1.2}$$

where we sum over all the particles the product of the ith particle mass m_i and the square of its distance from the axis of rotation r_i^2,

and
$$\omega = \lim_{t \to 0} \left(\frac{\Delta\theta}{\Delta t}\right) \tag{A.1.3}$$

θ being the angle in radians through which the particles rotate. For one particle in a circular orbit, simple relationships exist for these quantities, namely:

$$I = mr^2 \tag{A.1.4}$$

and
$$\omega = \frac{v}{r}. \tag{A.1.5}$$

Thus the angular momentum of an electron in circular orbit about the nucleus is

$$L_e = m_e vr \tag{A.1.6}$$

and together with Bohr's quantization condition

$$L_e = n\hbar \tag{A.1.7}$$

leads to
$$m_e vr = n\hbar\ . \tag{A.1.8}$$

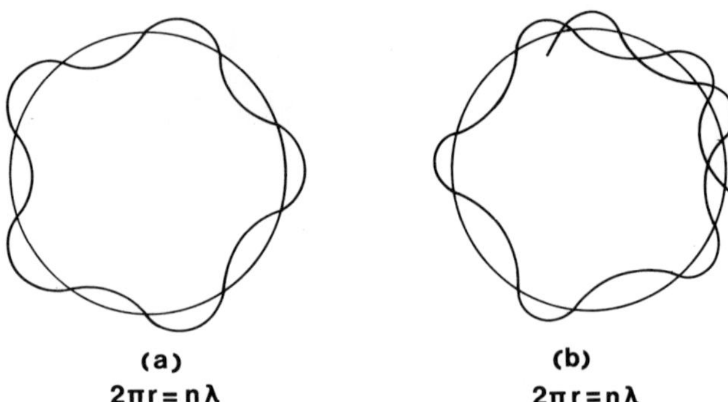

FIGURE A1.1. (a) A standing wave about a circular orbit. (b) Not a standing wave.

Another way of visualizing the quantization of the electron in the Bohr atom occurs if we consider the wave nature of the electron in a circular orbit. We assume that for a stationary state of the hydrogen atom, the wave associated with the electron's orbit must be a standing wave; otherwise it will destructively interfere. For this to happen there must be an integral number of wavelengths around the circular orbit as depicted in Figure A.1.1(a). This requires the relationship

$$2\pi r = n\lambda, \tag{A.1.9}$$

where r is the radius, λ the wavelength, and n an integer.

Now applying the deBroglie relation we have

$$2\pi r = n\lambda = \frac{nh}{p} = \frac{nh}{mv} \tag{A.1.10}$$

Then remembering Equation A.1.6 for a circular orbit, we again arrive at the Bohr quantization condition,

$$L_e = n\hbar \tag{A.1.11}$$

Appendix II
Spherical Polar Coordinates

Orbitals are generally depicted as contour diagrams showing the angular dependence of the wave functions. These angular dependences are best seen in a spherical polar space, which we now define.

From an origin O in Figure A.2.1 we locate a point P in space by Cartesian coordinates x_1, y_1, and z_1. We can equally well define the point P in terms of an angle ϕ between the POZ and XOZ planes, an angle θ between the lines OP and OZ, and r, the distance from the origin to the point P.

The rectangular Cartesian coordinates for the point P in terms of these new coordinates are:

$$x = r \sin\theta \cos\phi$$
$$y = r \sin\theta \sin\phi$$
$$z = r \cos\theta.$$

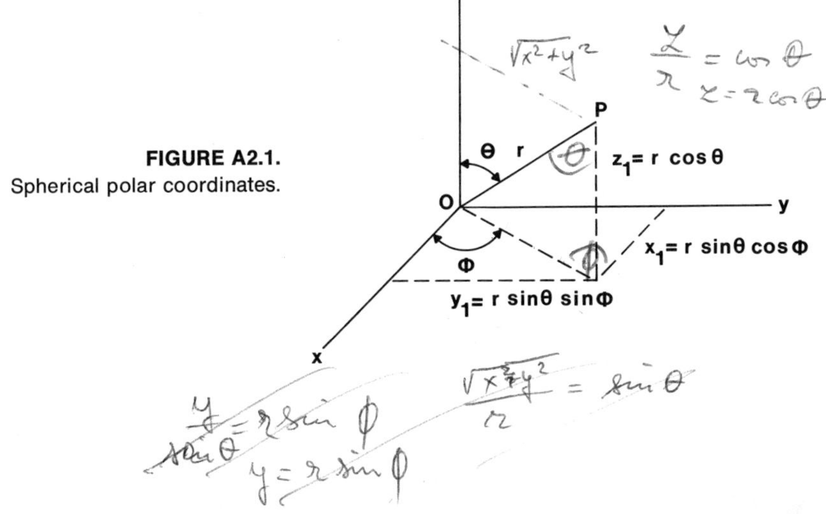

FIGURE A2.1.
Spherical polar coordinates.

The new coordinates r, θ, and ϕ are termed *spherical polar coordinates*. They in turn can be expressed in terms of the rectangular Cartesian coordinates:

$$r = \sqrt{x^2 + y^2 + z^2}$$

$$\theta = \text{arc cos} \frac{z}{(x^2 + y^2 + z^2)^{1/2}}$$

$$\phi = \text{arc tan } y/x.$$

Notice that the ranges of possible values are:

$$0 \leqslant r \leqslant \infty, \quad 0 \leqslant \theta \leqslant \pi, \quad 0 \leqslant \phi \leqslant 2\pi.$$

Appendix III
Atomic Orbitals and Magnetic Quantum Numbers

When the electronic structure of atoms is being discussed in simple terms, as in Chapter 3 (pp. 47–53) it is customary to assign specific magnetic quantum numbers to each orbital, for example, $m_l = +1, 0$, and -1 to the three 2p orbitals. The explicit mathematical functions representing these atomic orbitals for the hydrogen atom in spherical polar coordinates (Appendix II) are:

$$2p_0 = f(r) \cdot r \cos\theta$$

$$2p_{+1} = f(r) \cdot \frac{r}{\sqrt{2}} \sin\theta \, e^{i\phi}$$

$$2p_{-1} = f(r) \cdot \frac{r}{\sqrt{2}} \sin\theta \, e^{-i\phi}$$

where $$f(r) = \frac{1}{\sqrt{32\pi a_0^5}} \exp\left(\frac{-r}{2a_0}\right)$$

a_0 being the Bohr radius.

The functions corresponding to $2p_{+1}$ and $2p_{-1}$ are difficult to visualize and to use in qualitative discussions of atomic structure and valency problems because they are complex functions, involving $i = \sqrt{-1}$. However, the magnetic quantum number is a useful label only for discussing special magnetic and spectroscopic properties of atoms. We find it convenient to forego this label and to make our orbitals real. This is done by making use of a mathematical property of degenerate orbitals, namely, that we are allowed to make linear combinations of the degenerate orbital functions. If the combinations are linearly independent, they are equally good orbital functions for representing the electronic structure of the atom. We can thus seek combinations of $2p_{+1}$ and $2p_{-1}$ that are real functions and so easier to portray. The following combinations are convenient

$$\frac{1}{\sqrt{2}}(2p_{+1} + 2p_{-1})$$

$$-\frac{i}{\sqrt{2}}(2p_{+1} - 2p_{-1}).$$

Having relinquished m_l as a label we need some other means of identifying these new functions. As shown in Appendix II, the relationships between Cartesian and spherical polar coordinates are:

$$\begin{aligned} x &= r \sin\theta \cos\phi \\ y &= r \sin\theta \sin\phi \\ z &= r \cos\theta \end{aligned}$$

and so the above combinations are simply

$$\frac{1}{\sqrt{2}}(2p_{+1} + 2p_{-1}) = xf(r)$$
$$\frac{-i}{\sqrt{2}}(2p_{+1} - 2p_{-1}) = yf(r).$$

Furthermore:

$$2p_0 = zf(r).$$

Thus we logically label the two new functions and relabel $2p_0$ as follows:

$$\begin{aligned} 2p_x &= xf(r) \\ 2p_y &= yf(r) \\ 2p_z &= zf(r). \end{aligned}$$

The situation is analogous for 3d orbitals, the functions for $m_l = \pm 1$ and $m_l = \pm 2$ being complex functions. In this case the combinations that give the most convenient real orbitals are:

$$3d_{z^2} = 3d_0 = \frac{1}{\sqrt{3}}(3z^2 - r^2)F(r)$$

$$3d_{xz} = \frac{1}{\sqrt{2}}(3d_{+1} + 3d_{-1}) = 2xzF(r)$$

$$3d_{yz} = -\frac{i}{\sqrt{2}}(3d_{+1} - 3d_{-1}) = 2yz\,F(r)$$

$$3d_{x^2-y^2} = \frac{1}{\sqrt{2}}(3d_{+2} + 3d_{-2}) = (x^2 - y^2)\,F(r)$$

$$3d_{xy} = -\frac{i}{\sqrt{2}}(3d_{+2} - 3d_{-2}) = 2xyF(r).$$

Notice that these five real d-orbitals are not spatially equivalent as were the three p-orbitals: for instance, the $3d_{z^2}$ orbital has an angular dependence quite unlike the other d-orbitals (see Figure 3.21). It is possible to construct a set of five equivalent d-orbitals if desired [see *J. Chem. Educ.* (1968) *45,* 45; (1970) *47,* 15]. These five equivalent d-orbitals have an angular dependence that somewhat resembles a d_{z^2} orbital, but they point to the corners of a pentagonal antiprism.

Since much of our discussion of valency involves the directional properties of bonds, orbitals pointing to the corners of a pentagonal antiprism are of little use because this geometrical form rarely occurs in molecular structures. Thus we exclusively use the $3d_{z^2}$, $3d_{xz}$, $3d_{yz}$, $3d_{x^2-y^2}$, and $3d_{xy}$ orbitals as our real orbital functions.

Appendix IV
Calculations on CaCl, $CaCl_2$, and $CaCl_3$

We consider three possible ionic combinations of gaseous atomic Ca and gaseous atomic Cl:

1. CaCl:

   The processes involved follow that given for NaCl in Chapter 5 (pp. 89–93).

   a. Conversion of Ca and Cl into the ions Ca^+ and Cl^-:

   $$\Delta E_a = I_1(\text{Ca}) - A(\text{Cl}) = 6.1 - 3.7 = 2.4\ \text{eV}.$$

   b. Combinations of pairs of ions to form diatomic CaCl separated by 280 pm:

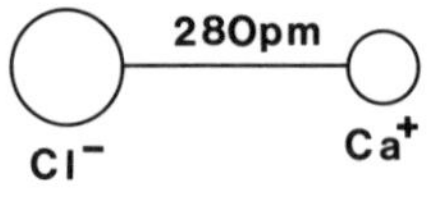

   280pm

   Cl^- Ca^+

   $$E_b = -\frac{1440}{280} = -5.1\ \text{eV}$$

   c. The overall energy change is:

   $$\Delta E = \Delta E_a + \Delta E_b = -2.7\ \text{eV}.$$

2. $CaCl_2$:

   a. Conversion of Ca into Ca^{2+} and 2Cl into $2Cl^-$:

   $$\begin{aligned}\Delta E_a &= I_1(\text{Ca}) + I_2(\text{Ca}) - 2A(\text{Cl}) \\ &= 6.1 + 11.9 - 2(3.7) = 10.6\ \text{eV}.\end{aligned}$$

   b. Combinations of Ca^{2+} with $2Cl^-$ to form linear triatomic $CaCl_2$ molecule again with a Ca–Cl internuclear separation of 280 pm.

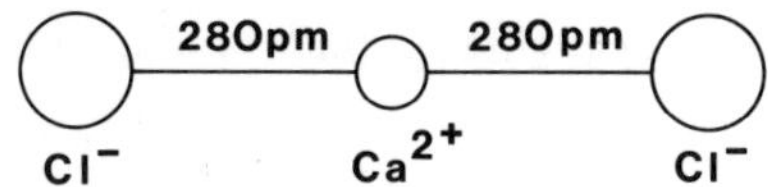

$$\Delta E_b = -\frac{2(1440)}{280} - \frac{2(1440)}{280} + \frac{1440}{560} = -18.0 \text{ eV}.$$

c. The overall energy change is

$$\Delta E = \Delta E_a + \Delta E_b = -7.4 \text{ eV}.$$

3. $CaCl_3$:

a. Conversion of Ca into Ca^{3+} and 3Cl into $3Cl^-$:

$$\begin{aligned}\Delta E_a &= I_1(\text{Ca}) + I_2(\text{Ca}) + I_3(\text{Ca}) - 3A(\text{Cl}) \\ &= 6.1 + 11.9 + 51.2 - 3(3.7) = 58.1 \text{ eV}.\end{aligned}$$

b. Combination of Ca^{3+} with $3Cl^-$ to form a tetratomic planar triangular $CaCl_3$, again with an internuclear separation of 280 pm:

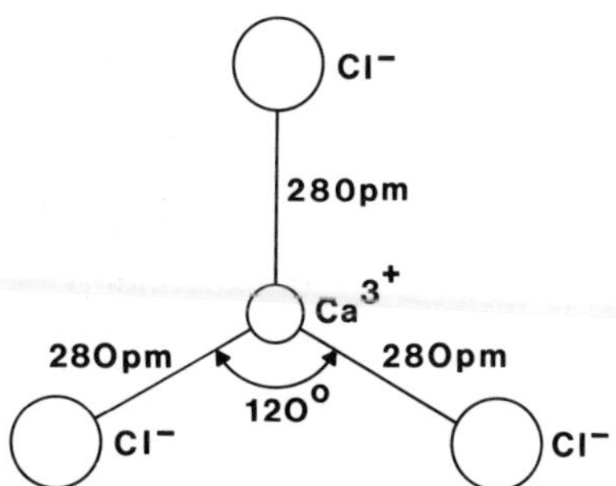

$$E_b = -\frac{3(1440)}{280} - \frac{3(1440)}{280} - \frac{3(1440)}{280} + \frac{3(1440)}{\sqrt{3}(280)} = -37.4 \text{ eV}.$$

c. The overall energy change is:

$$\Delta E = \Delta E_a + \Delta E_b = +20.7 \text{ eV}.$$

Since process (2) leads to the lowest potential energy [−7.4 eV vs. −2.7 eV for process (1)], it has the greatest stability. Hence we might conclude that $CaCl_2$ is the preferred way of chemically combining calcium and chlorine. Before making a hasty decision, there are two additional factors to consider.

Firstly, we have used the internuclear separation of 280

pm throughout the calculations. This is the value determined experimentally for $CaCl_2$. For CaCl the distance should be greater, because the Ca^+ radius will be greater than Ca^{2+}. This factor works to the disadvantage of CaCl in terms of stability, but in the case of $CaCl_3$, the radius of Ca^{3+} should be smaller than Ca^{2+}. This should make $CaCl_3$ more stable electrostatically than we have calculated. Will it be enough to render $CaCl_3$ the preferred form of combining calcium and chlorine?

Unfortunately, we have no experimental information of the size of Ca^{3+}, but to solve this dilemma, let us consider the extreme case of zero radius of Ca^{3+}. The interionic distance would then be just the radius of Cl^- (viz., 181 pm). We then have

$$\Delta E_b = -\frac{9(1440)}{181} + \frac{3(1440)}{\sqrt{3}(181)} = -57.8 \text{ eV}$$

hence

$$\Delta E = \Delta E_a + \Delta E_b = 58.1 - 57.8 = +0.3 \text{ eV}$$

so that the overall energy change, even for this extreme geometry does not lead to stability.

Notice that the dominating factor for the $CaCl_3$ calculation is the large value, 51.2 eV, of the third ionization potential of calcium. This renders $CaCl_3$ unfavorable, despite the favorable effect of the +3 ionic charge on calcium. For $CaCl_2$, because the second ionization potential of calcium is only approximately twice the first, the unfavorable change in ΔE_a is more than compensated by the more favorable value of ΔE_b, the latter stemming from the +2 ionic charge on calcium.

The second complicating factor is that more extensive clustering of ions can lead to more favorable potential-energy arrangements than those that we have just computed for simple diatomic, triatomic, and tetratomic molecules. Just as for NaCl in Chapter 5 (pp. 94–96) we find that the most stable form of $CaCl_2$ is a large stack of close-packed ions in the form of an ionic crystal.

Index

C

D

E

F

G

H

M

N

Valence Bond } Theory
M.O.